CRUSHED

CRUSHED

HOW A CHANGING CLIMATE IS ALTERING THE WAY WE DRINK

BRIAN FREEDMAN

ROWMAN & LITTLEFIELD
Lanham • Boulder • New York • London

Published by Rowman & Littlefield
An imprint of The Rowman & Littlefield Publishing Group, Inc.
4501 Forbes Boulevard, Suite 200, Lanham, Maryland 20706
www.rowman.com

86-90 Paul Street, London EC2A 4NE

Distributed by NATIONAL BOOK NETWORK

British Library Cataloguing in Publication Information Available

Library of Congress Cataloging-in-Publication Data is Available

ISBN 978-1-5381-6630-7 (cloth)
ISBN 978-1-5381-6631-4 (ebook)

For Steffi, Sophie, and Olivia,
who inspire and amaze me every day.

And for my parents, Linda and Arnie,
who never stopped believing in me.

CONTENTS

Acknowledgments ix

Introduction: Smoke Gets in Your Eyes xiii

1 The Scorching of California's Wine Country 1

2 Waking Up to Climate Change 17

3 Desert Roots 39

4 A Bright Future for Overcast England 63

5 American Spirit 87

6 What Happens in the South When Temperatures Go North 111

7 Deep Freeze in Hill Country 133

8 The Philosopher-Farmer of the Western Cape 155

Notes 179

Bibliography 185

Index 191

About the Author 193

ACKNOWLEDGMENTS

This book never would have come to fruition without the tireless work of my amazing agent, Sarah Phair, of Sanford J. Greenburger Literary Agents: her guidance, advice, and support have been invaluable. I also owe a huge debt of gratitude to my excellent editor at Rowman & Littlefield, Suzanne Staszak-Silva, who saw the importance of this topic and brought the book to life. Naomi Minkoff did a fantastic job copyediting the manuscript, and Alyssa Henkin got the ball rolling years before I ever wrote a word of this. Thanks as well to Erin Sinesky Lovett, whose enthusiasm and guidance are very much appreciated.

Kristen Green-Kutch, Angela Duerr of Cultured Vine, Jennifer Scott of SMWE, Amy Mironov, Christina Seillan, Josh Greenstein of Royal Wine Corp., Erik Segelbaum, the IWPA, Peter Breslow, Mary Collum, Hamdy Khalil, Caroline Paulus, Wagstaff Media + Marketing, Amy Preske, Christi Crosby, Buffalo Trace, Sazerac, Eva Valdebenito, Nora Favelukes, Todd Nelson, Denise Clarke and Texas Fine Wine, Jim Clarke and Wines of South Africa, Aaron Meeker, Erick van Zyl and South African Tourism, the teams at Colangelo & Partners, Jarvis Communications, Nike Communications, J.A.M. PR, Sopexa, LVMH, Serious Business PR, Lisa Shea, Casey Shaughnessy,

Claire Hennessy, Alex Schrecengost, Wilson Daniels, Lisa Mattson and the team at Jordan, Zach Groth, Laura Baddish, Evins Communications, Savona Communications, Sazerac, Jackson Family Wines, Teresa Wall and Napa Valley Vintners, Beth Cotenoff and Teuwen Communications, Andrea Duvall and Brown-Forman, KLGPR, Alex Koiransky and the team at FAIR Spirits, Courtney Schiessl Magrini and *SevenFifty Daily*, and the kind, smart, generous colleagues and amazing PR and marketing teams and regional and national wine and spirit boards that I've had the privilege of traveling with over the years: Thank you all. Massive thanks as well to the wineries and distilleries, many of them not in this book by name, whose expertise and open doors have laid the foundation for it.

Very warm thanks also go out to two mentors and friends: John Mariani, who gave me my first big break, and Ray Isle, who brought me on as a contributor to *Food & Wine* digital. Both have been beyond kind, generous, and helpful. Sean Flynn, my excellent editor at F&W digital. Lyn Caum and Toby Thompson, who did so much to nurture my love of writing. I'm grateful as well to the editorial teams at Forbes.com and F&W digital; Melanie Schwenk at *Whisky Advocate*; Gerard Holden of Holden Manz Wine Estate; Ian Freedman and Paul Balbresky; Jordan Balbresky, with whom I used to dream of being a writer one day over way too many martinis in New York twenty years ago; Jessica Dupuy, Jonathan Cristaldi, and Matt Ulmer, amazing writers, friends, and professional sounding boards; Scot "Zippy" Ziskind and the Dead Guys Wine Society; the Wine Media Guild; and so many more—especially the people who are featured in this book, who were so generous with their time and trusting in my ability to tell their stories. I hope I've done right by them.

Lots of love and gratitude to my sister Amy Shubert, who has spent the past two years dealing with unimaginable challenges as a school nurse during a pandemic: she's always believed in me, and was an early reader of my first chapter, for which I'm eternally grateful. And to my brother-in-law Mike Shubert—whiskey lover, meat smoker extraordinaire, and all-around great guy.

They may be gone, but I think about my Nana Gwen Balbresky and Pop-Pop Charlie Freedman every day—their joy and zest for life are in this book.

I am deeply indebted to my parents, Linda and Arnie Freedman, whose love, support, enthusiasm, and warmth mean the world to me. The love of food, wine, and dinner-table conversation that they instilled is why I'm in this line of work in the first place.

Olivia and Sophie—the "Freedman Ladies," two of the most amazing people I know. It's a joy and a privilege watching them grow up. I have no idea what we did to deserve them, but Steffi and I are the luckiest people in the world to raise daughters like these two.

And finally, absolutely none of this would have been possible without my wife, Steffi—this book is as much hers as it is mine. She is the brains behind the operation, the extraordinary human who keeps our crazy household running, proofreader, sounding board, advice giver, exemplary mother, incredible wife, best friend, and soul mate. I'm hereby officially ignoring Larry David's hesitations and agreeing, happily, to eternity.

(And as promised, a big thanks to Murray and Gianluca too.)

INTRODUCTION

Smoke Gets in Your Eyes

Something was wrong with the electricity, and I couldn't figure out what it was. Or rather, the problem was clear—there was no power in my room—but at the time, I had no idea what had caused it. This was Indian Springs Calistoga, after all, the destination hotel and spa in the northern reaches of Napa Valley. Room rates can climb well into car-payment territory, and given its reputation for high-end facilities and service, the lack of power made no sense. At the front desk, I encountered a small group of other grumpy, confused guests, all of us in various shades of dishevelment. "The only information we have," the flustered attendant told us, "is that PG&E had to turn off the power to the area." At the time, we didn't know why they'd done that or how long the lack of electricity might last. We all trudged off to our rooms, passing other guests making the same pilgrimage to the front-of-house team in search of answers.

I had spent the previous couple of days in Napa, and the trip had been full of the usual perks that my line of work provides with almost embarrassing frequency: meals that stretch on for hours, wines that I couldn't afford even if I wanted to spend the money on them, time wandering around some of the most beautiful landscapes in the world.

Now, however, it was clear that something was terribly wrong. A little before lunchtime, word had begun spreading that a fire exploded earlier in the morning, forcing Pacific Gas & Electric to cut off power out of an abundance of caution. As afternoon bled into evening, my eyes began to burn and water in the increasingly smoke-tinged air. My red-eye flight took off late that night, depositing me back in Philadelphia the next morning to news that the Camp Fire was burning a path of unprecedented destruction.

By the time it was finally contained, the fire had torched 153,336 acres, according to Cal Fire. It had also destroyed 18,804 structures and killed eighty-five people.[1] As I write this in 2021 and 2022, it still ranks as the most destructive fire in the recorded history of California.

I didn't know it at the time, but that experience—of being there at the beginning of a blaze of such incomprehensible magnitude, breathing in its smoke, complaining to the front-desk clerk of a luxury hotel that my electricity wasn't working when less than three hours away people were fighting to save their homes and their lives—was the kernel of this project. As a wine, spirit, travel, and food writer, I've seen and tasted firsthand how climate change is impacting our world. It's virtually impossible these days to sit down for a drink with a winemaker or distiller without the conversation, at some point, turning to the unprecedented challenges they're facing, whether it's heat waves, too much or not enough rain, the invasion of insects and plant viruses that had never been present in their part of the world before—it can all feel fairly biblical and more than a bit overwhelming.

That is why I wrote this book, to better understand the ways in which climate change is affecting the people who are responsible for coaxing the grapes and the grains to maturity and the others who then turn those raw materials into the wines and spirits that are so easy to take for granted. Climate change, after all, is affecting the world of wine and spirit production in countless familiar and unprecedented ways, but it's all too easy to forget about all of that once the cork is popped and the glasses are filled.

In researching and writing this book, I chose to focus on eight distinct parts of the world and tell the stories of the growers and producers in each that are not just having to deal in increasingly desperate ways with the effects of the changing climate but in some cases, struggling for their very professional survival. The speed with which climate change is impacting environments around the world is far greater than experts predicted even just a decade ago, and the people profiled in this book are, unwittingly, at the front lines of a battle that's really just begun in earnest.

The Sunday night before I finished my draft manuscript of this book, *60 Minutes* aired a double-length segment on the ways in which climate change is affecting wine production around the world. As my wife and I enjoyed a low-key dinner at a local Italian BYOB, my phone began buzzing before we had finished the fried calamari and continued vibrating in my pocket all the way through the pasta course. Did I know, friends texted me, how bad it was in the world of wine? That Champagne had just completed its smallest harvest in nearly half a century? That Burgundy is facing a potentially serious shortage? *Funny you should ask* . . . I replied.

This isn't a book of unrelenting struggle, however. Over the years, I've learned that the people who choose to work in wine and spirit production are often some of the smartest and most resourceful in the world. Many of them are in this book: the physician-turned-winemaker who is completely changing the ways in which wine is made in Texas Hill Country; the agronomist in Israel who is bringing back to vineyards plants, animals, and insects that had been separated from their natural ecosystems for too long; the farmer-philosopher in South Africa working to tie together the health of the land and the potential for a better life among a post-apartheid generation that has been denied those opportunities for too long.

There are also places in the world that are benefiting from the effects—at least for now—of a changing climate. The sparkling wine producers of southern England, for example, are riding a wave of unprecedented popularity and achievement, much of it thanks to warming temperatures. Space, time, and COVID-related difficulties in traveling meant that some of the places I wanted to visit weren't able to be included in this book. But that doesn't mean that their stories are any less important or any less illuminating. In Denmark, for example, temperatures are warming up enough that a number of brave, pioneering professionals are successfully growing wine grapes and vinifying them into wine.

I could have written an entire book on just the technological advances that are being made to help growers and producers survive and thrive in this dramatically changing world. Bespoken Spirits, which I include in my chapter on American whiskey, is leveraging machine learning and artificial intelligence to shorten the amount of time a whiskey has to age from years to days. Or EarthOptics, an agricultural-technology start-up that aims to codify and standardize food and beverage labels to clearly convey how much carbon dioxide was sequestered in the soil over the course of the production of that particular food or drink item. Or the glass and cork manufacturers that are changing the ways in which wines and spirits are packaged, making them lighter, more sustainable, and less damaging to the environment. For all of the struggle and destruction caused by climate change, it is spurring on a level of innovation that is unprecedented in the world of wines and spirits.

Many large companies, perhaps counter to how they're often discussed, are taking a remarkable and laudable lead. Among many consumers and a small but vocal segment of beverage professionals, there is a tendency to fetishize the climate change–combating efforts of smaller companies while having a knee-jerk reaction against larger ones. And while there have historically been plenty of questionable and destructive decisions that the biggest players have made, there's been a real turnaround in recent years, which is having a tremendous impact. The largest producers, after all, can have an outsized effect on climate change and how we handle it, for better or worse, by virtue of both their sheer size and the trickle-down effect that often flows from the biggest to the smallest companies (things can certainly move in the other direction too).

Right before the pandemic put a halt to such trips, for example, Moët Hennessy USA brought me to Paris to attend Vinexpo 2020 in February of that year, with a specific focus on a three-day conference-within-a-conference that they hosted dealing with sustainability in the world of wines and spirits. Speakers ranged from winemakers to university professors to heads of corporations, and the discussion was eye opening and left me with a deep sense of optimism about the future of my industry's efforts to learn more about and combat climate change. Jackson Family Wines, over the course of the pandemic, hosted a series of Zoom discussions called "Rooted for Good," which explored both climate change and social responsibility as well as the ways in which wine professionals can positively impact both. Spirit producers all over the world are also doing their part to change the trajectory of the issues we're facing. Tequila Cazadores, for example, which is owned by Bacardi, generates over 99 percent of its energy from renewable sources, including solar and wind. Patrón creates fertilizer from its spent agave fibers, as well as the fibers of ten other local distilleries, to help nourish their agave plantings in a more sustainable manner.

The list of efforts by producers both large and small could go on for hundreds of pages. The important point to note is that these initiatives are beneficial from an ethical and a marketing standpoint (conscientious consumers are increasingly more likely to buy from environmentally responsible producers) as well as a business one. The process of growing grapes, agave, grain, and the rest and producing wines and spirits from those raw materials is big business, and players at every point along the production continuum increasingly understand that the old ways of doing things simply won't afford them a bright and financially viable future. They also tend to realize that even if the investment required to increase sustainability and social responsibility is significant, it will benefit the bottom line in the long term. Wheyward Spirit, a new spirit company in Oregon, uses what would have been cast-off whey from cheese production to craft a clear spirit of depth, character, and fantastic versatility behind the bar. FAIR Spirits, with which I traveled to Bolivia in 2019 to visit some of the villages that grow the Fair Trade quinoa that's used in their vodka, is working to expand their footprint around the world.

In the United States alone, the wine and spirit industries at wholesale accounted for more than $120 billion in 2021, according to IBISWorld—and that number is growing each year.[2] Globally, reports Yahoo Finance, they cleared *half a trillion dollars* in 2021 and are expected to hit over $735 billion by 2025.[3] This means that a massive swath of people and businesses connected to the beverage-alcohol industry around the world are at risk from climate change. It also, fortunately, means that the impetus and resources exist to find solutions to these problems. Even eventual savings of a few percentage points per year will be a massive number.

The beverage-alcohol industry is stepping up in important and often fantastically forward-thinking ways. No book can cover all of them, just as no single work can exhaustively explore every producer and every region around the world being affected by climate change. The aim of this book, however, is to pull back the curtain and show readers what's happening behind the scenes in an industry that's not often part of the popular climate-change conversation. Too often, we focus on how a warming world affects food production, without looking at the impacts on an industry that is responsible for hundreds of billions of dollars in revenue and millions of jobs . . . not to mention the joy and sense of sanity that a good drink can bring in challenging times, as so many of us have learned during these two years (and counting) of COVID.

My hope is that *Crushed* shines a light on the climate challenges that wine and spirit producers around the world are facing with greater and more urgent frequency and showcases the ingenuity, creativity, and passion that the best of them are bringing to their efforts to not just survive in this brave new world but to thrive.

I raise a glass to their efforts every day.

Brian Freedman
Ardmore, PA
December 2021

1

THE SCORCHING OF CALIFORNIA'S WINE COUNTRY

Jamie Kutch was ready for a break. Every year around this time, between the weeks leading up to harvest and that seemingly impossible moment when all the wines are safely in their barrels, he sees his family far less than he'd like, and decent REM-cycle sleep is nothing more than a hypothetical. This year—2017—had been no different. But now it was October, and after a months-long sprint and more than a few restless nights, he and his wife, Kristen, a wine publicist, and their five-year-old son, Clayton, had finally unplugged.

Their hotel room in Half Moon Bay looked out over the cliffs that drop off into the thrumming Pacific below. The juxtaposition of the wildness of the Northern California coastline and the well-appointed room itself—spacious bathroom, all the soaps and lotions they could ever need, robes plush enough for Clayton to get lost in—was exactly what the family needed. Kristen and Clayton arrived the day before, but their getaway didn't officially begin until they were all together again and the main hurdle of the vintage was behind them.

They spent the morning exploring the nearby Purisima Creek Redwoods Open Space Preserve, all of them struggling to comprehend the sheer size of the ancient trees, some more than thirty stories high. Later on, they splashed in the pool, wandered around downtown, and reveled in finally getting to spend time together as a family, which they hadn't done much of since late summer. Clayton passed out as soon as they got back to their room; Jamie and Kristen took advantage of the time alone to pop open a celebratory bottle of white Burgundy and enjoy the view from their terrace.

I needed this so badly, Jamie thought, finally slipping into vacation mode and breathing the salt air breezing in from the ocean. He took a sip of Chardonnay. It was the Coche-Dury Meursault, one of the more approachable wines from a producer whose best bottlings cost thousands of dollars, a perfect way to mark what was looking to be yet another successful vintage for a winemaker who has always referred to the great Chardonnays and Pinot Noirs of Burgundy for inspiration.

Domaine Jean-François Coche-Dury traces its roots back a little less than half a century, to its founding in the mid-1970s—the blink of an eye when compared to other legends of the region. Domaine de la Romanée-Conti (its most ardent collectors generally refer to it as DRC, for short), for example, was born in 1232, when the Abbey of Saint-Vivant acquired just under four-and-a-half acres of vineyard land in Burgundy. Yet despite Coche-Dury's relative youth, its reputation is massive; the vines it grows on its approximately twenty-two acres of vineyard land—most of it divvied into relatively small individual plots—produce some of the greatest Chardonnay on the planet. Competition for the vanishingly small number of bottles is so fierce that collectors pay thousands of dollars in even less-than-stellar vintages for its flagship, the Corton-Charlemagne Grand Cru, which it grows on less than an acre of land. The Meursault, on the other hand, can be found for well less than $1,000—still serious money, but at least approachable. And for a winemaker like Jamie Kutch, it represents everything that makes Burgundy such an inspiration, the ways in which each village, each parcel of vines, produce wines of such distinct and often idiosyncratic character.

That's the nature of great Chardonnay and Pinot Noir; they respond to seemingly every variation in soil, microclimate, and geology in a way that their most passionate fans can detect by sniffing and sipping the liquid in the glass. Coche-Dury's Corton-Charlemagne costs more than its Puligny-Montrachet Les Enseignères, and DRC's Pinot Noir from its Romanée-Conti vineyard costs so much more than its Pinot from the Echézeaux vineyard because both the Corton-Charlemagne and the Romanée-Conti are, as far as wine can ever be objectively qualified, more complex and complete wines. Those highly venerated vineyards produce wine of such a unique character because over the past hundreds of millions of years, everything, from the movement of Ice-Age glaciers to deposits on the bed of the ancient ocean that once covered the area, has resulted in a patch of the planet that is particularly well suited to Chardonnay and Pinot Noir. This concept has always appealed to Jamie, and 5,700 miles away, he strives to express the uniqueness of the American vineyards where his grapes are grown.

He topped up their glasses and was just sloughing off the last of the harvest season's stress when both of their phones started buzzing incessantly.

Oh, shit . . . Jamie recalled thinking, staring at his screen in disbelief. "Kristen?"

"Yeah, I'm seeing it too. What's happening at the winery?"

"I have no idea, and no one's responding either."

The fire that began the night before, on October 8th, had been spreading out of control, whipped by fierce winds and fueled by historic drought. But even as news reports kept their phones pinging all day, it didn't seem possible that it would pose a serious threat to his winery in Sonoma. Now, however, the story told by the satellite images and footage from the ground was incontrovertible; his family's entire future was at risk.

Jamie had to get to the winery, and in a matter of minutes, he and Kristen were up, throwing their clothes into their bags, tossing Clayton's gear into his own duffel, and heading through the door. Though his increasingly buzzed-about wines are crafted primarily from individual vineyards dotting the remote Sonoma Coast, the winery itself is located on East Eighth Street in Sonoma. As far as he could tell, the flames were getting uncomfortably close.

Five minutes later, they were in their cars, making their way to the 101 in the hopes of finding the winery, and the wine inside it, still intact. Both of them were obsessively checking their phones as they drove, texting and calling everyone they could in an effort to gather as much information about the fire as possible. Already, the sky above the Kutches's cars was turning an oddly vivid shade of orange; the smoke from the fire had traveled far enough that it had begun diffusing the light in menacing—yet oddly beautiful—ways.

As they sped back home, their wineglasses remained untouched on the table between their two chairs on the balcony, a few sips remaining in each. Coche-Dury's lovingly crafted Chardonnay—one of the most clamored for in the world, and for Jamie and Kristen the symbol of yet another successful vintage for Kutch Wines—was slowly evaporating inside.

Jamie and Kristen met in New York City in the early 2000s, when she was a publicist just beginning to make a name for herself and he toiled away as a NASDAQ trader at Merrill Lynch. Like many of the men in the office—and they were almost all men back then, with suit shoulders broad enough to make even the skinniest among them look like comic book superheroes—Jamie developed a passion for wine. Dinners at Gotham Bar and Grill were punctuated by bottles of Burgundy. Gramercy Tavern, with a pre-TV Tom Colicchio in the kitchen, was where Jamie and his friends discovered the cult bottlings of California. Strip steak and a mortgage-payment bottle of Cabernet? Yes, please! Magnums of Champagne were popped at Moomba, and Château Margaux sloshed around in glasses high above it all at Windows on the World.

But unlike most of the budding oenophiles in the office, many of whom used wine as a way to show off their increasingly impressive paychecks and annual bonuses, Jamie found himself falling down the rabbit hole of wine production. How, he wondered, could something as simple as grape juice channel the place where the fruit was grown with such precision and emotion? Why did a flute of Champagne Salon or a glass of Pinot from Burgundy's Richebourg vineyard fascinate him so much?

Jamie would go home after those boozy nights, and instead of popping a few Advil and passing out next to Kristen, he'd poke around the early wine-focused message boards, absorbing as much information and knowledge as he could. He would spend hours arguing the relative merits of clay soil versus sand and whether gravity-flow wineries really made all that much difference. Eventually, perhaps inevitably,

Jamie started to experiment with winemaking himself in his small downtown apartment. Unexpectedly enough, the wine wasn't all that bad. Nothing that would ever cause any particular sense of fear in the gilded halls of Château Lafite, but good enough that he started sharing bottles with some of the guys at work.

He eventually got up the courage to email his California wine heroes. A decent number of them agreed to taste his wine and give him their honest feedback . . . which mostly consisted of surprise and support. Sending a professional winemaker samples of homemade red or white is like emailing Jay-Z the rap demo you recorded on iTunes—utter insanity and an invitation for silence at best and flat-out rejection at worst. But Michael Browne, a California Pinot Noir visionary who would later sell his Kosta Browne label to the venerable Duckhorn Wine Co., tasted enough potential that he suggested Jamie take some time off from finance to work a harvest and see how the pros did things.

A year later in 2005, Jamie and Kristen packed up their apartments, walked away from their lucrative careers, and decided to make a go of it in wine country. Their friends were incredulous. Some of his colleagues openly called Jamie's actions "insane," often after their third or fourth glass of $1,000 Harlan Estate Cabernet—which, they were always sure to point out, Jamie would never be able to afford without his Merrill Lynch paycheck. But when they weighed the very real costs against all the other benefits, there was no question: California was the only acceptable choice, especially since they wanted to have a child one day. Their new lifestyle would be more in line with what they envisioned for themselves and their eventual family. The balance would be better, and the risk of failure was more than worth the potential benefits of success.

Jamie's first vintage was met with the same silence that most new wines are greeted with. The wine world, after all, is like the restaurant business; the supposed glamour and lifestyle make it appealing to a far wider swath of the population than could actually pull it off . . . or want to. The hours are grinding, the margins are slim, and financial ruin is often just a single bad vintage away. A winemaker once lamented to me how jealous she was of chefs.

"I mean, they have a bad night? Sucks for you," she said. "But there's always tomorrow, and the next day and the next to make it better. As a winemaker, I get one chance a year. So over an entire career, that's thirty or forty harvests? And if I screw up one of them, or nature takes a shit on us that year and there's no good fruit, or rain right before the harvest and the grapes get all swollen with water and the juice loses concentration, or there's a fire or an earthquake—what do I do then?" For most winemakers, it's a struggle that lasts an entire career—assuming they ever make wine that consumers want to buy in the first place.

Weather, soil, the changing climate, and countless other variables of nature play a huge role in the quality and character of the grapes at harvest. But wine isn't just about the land itself. It's also a result of the winemaker's ability to express the vineyard and vintage conditions as well as his or her own vision for how best to achieve that. In the winery, decisions about how long to ferment the juice, what type of yeast to use, whether to employ oak barrels, and more impact the wine in the glass.

Starting in the 1980s, a critic named Robert Parker became the most influential and important wine writer in the world. His awarding of high scores to wines of power and concentration changed the way wine was made from Napa to Bordeaux and beyond. His perceived personal preference for so-called big wines—powerful reds, lush and oaky whites—led to an era of wines made to suit the style that he popularized. As a result, grapes were often picked later to impart more ripe-fruit flavors to the finished wine. Alcohol levels climbed as well, and the use of new oak barrels gave the wines in which they were aged distinct sweet-spice notes like vanilla and cinnamon. Many of California's most successful producers began crafting wine in this style, and the high scores they were awarded led to skyrocketing sales and prices. Nature may not have taken a backseat in this brave new world of wine, but it arguably wasn't the driving force anymore. The winemaker was now the star of the proverbial show.

When Jamie produced his first vintage, he didn't know how to make the more delicate, Burgundy-style wines he wanted to. As a new winemaker without the confidence to break with the accepted wisdom of the time, he followed what his mentors did when it came to deciding on the exact moment to harvest and how to handle the grapes once they arrived at the winery. On top of that, he didn't own any land. His grapes came from small parcels he leased in larger vineyards. It's a common arrangement, and one that tends to benefit everyone. Winemakers generally decide how they'd like their rows of vines to be farmed, and the vineyard managers provide advice and implement the agreed-upon plans. Winemakers who want to produce tens of thousands of cases of low-priced wine, for example, tend to want their grapes grown differently from those who craft small amounts of organic wine. Winemakers like Jamie, whose goals are to express the specific vineyard each of their wines are grown in, visit regularly, often insist on sustainable farming methods, and work closely with the vineyard managers. It's a collaborative effort, all with the end goal of providing the winemaker with the exact kind of grapes that he or she wants.

As a young winemaker new to the Sonoma scene, however, Jamie lacked the confidence to push against the advice of the vineyard managers. For their part, harvesting riper fruit meant a better chance at high scores from the critics, which translated to an ability to lease their land for more money or to earn higher prices for the grapes that they sold directly to various wineries around the region. The extra-ripe fruit he was sold that first vintage resulted in wines that were technically sound but that just didn't suit his palate. He fell in love with elegant, subtle Pinot Noir all those years ago in New York, but his first batch in California was weighed down with more alcohol than he wanted. The fruit tasted stewed too. These first wines were the opposite of what Jamie wanted to produce.

The next vintage, still searching for how to achieve the style he desired, Jamie decided to adjust his pick date, and he had his fruit harvested thirty days earlier than the year before. He was openly mocked by some of his winemaking heroes and asked whether he was trying to make sparkling wine, the grapes for which are picked far earlier than the ones for red.

Yet he persisted. His 2007 wine was lighter in body—earlier-picked grapes have more acidity and less color in the skins—all of which was a step in the right direction, but the wine still wasn't quite there. Jamie loved the character, the idea of it, but it didn't have enough glycerin or alcohol, which meant not enough texture and weight on the palate. The wines felt thin, and Jamie had to figure out from where his wine was going to get power, weight, and appropriate tannic backbone if his early harvesting took several of the main means of achieving them off the table.

This is when the idea of stem inclusion started to germinate. Stems, after all, are a significant source of tannins. Many winemakers de-stem their grapes prior to crushing them for fear of their wines coming off as "stemmy," which means they seem oddly astringent and vegetal when sipped. This sensation is similar to that of swishing a big swig of oversteeped iced tea around your mouth. Stem inclusion, however, has been an important winemaker's trick in Burgundy for centuries to help build greater structure into wines that otherwise might have lacked it. Jamie decided to include stems in his winemaking, which resulted in a Pinot with the structure he desired. The tannins present in the stems added a sense of concentration to every sip as well as aging potential for the bottles themselves. He was getting closer, but other problems persisted. The stems worked to build greater structure, but they took his wine too far in the other direction. Each sip wasn't quite silky enough for his taste. He was trying to make more structured wines, but these were verging on austere. How could he remedy that?

In 2008, he didn't have the chance to find out; smoke from the Lightning Complex fires, which rolled through the Anderson Valley, tainted Jamie's juice. Jon Bonné, former wine columnist for the *San Francisco Chronicle*, wrote that "some wines . . . were saddled with an ashy, bitter aftertaste not entirely unlike gargling the remnants of an ashtray."[1] Jamie was forced to sell off half of his production—a major financial hit, and one that put make-or-break pressure on him for the following vintage. Technology was available that some producers employed to mitigate the smokiness in their wines, but for Jamie, this was both too expensive and ran completely counter to what he believed wine should be. He didn't want to build a career as a chemist; he wanted to make wine that was expressive of the vintage and the place where the grapes were grown. And if the resulting bottles just didn't provide pleasure when the corks were popped, he'd take his losses, learn how to manage that risk in the future, and hopefully survive another vintage to make things right.

After 2008, Jamie continued harvesting his grapes earlier, but this time, for their own safety as opposed to stylistic reasons. Wine is at its most vulnerable before undergoing fermentation, when it's still in the stainless steel tanks in which the process occurs at a winery like Kutch's. Counterintuitively, even though oak barrels are prone to fire in a way that steel clearly is not, the liquid inside them, already fermented, is more protected on a chemical level than its unfermented predecessor. The sooner he could get his wine into barrels, he reasoned, the safer he'd be.

But that opened him up to a host of other issues. Grapes, like all fruit, start off acidic and only slowly gain sweetness as sugars develop and acids diminish. The best vintages are ones in which the diurnal shift—the difference between daytime highs and nighttime lows—is particularly pronounced, allowing for sugar accumulation during the day and acid retention at night. Pick too early, and the sugars won't be high enough. On top of that, there also won't be a long enough period of time between budbreak and harvest for each individual berry to develop the layers of complexity that translate into great wine. Jamie had to figure out how to start the growing season earlier for the vines he was working with.

After several years of trial and error, he and his vineyard partners hit upon a solution; by clipping the tip of the tendril that shoots skyward early in the spring, they could essentially send a message to the vine to redirect its energy to the fruit instead of the canopy of leaves that is its ordinary focus early in the season.

And so it went, vintage after vintage, a Socratic winemaking dialogue with himself, questions posed and then answered a full year later. It's been a long, arduous process, but with each passing year, he has dialed in on the techniques and decisions that get him closer to crafting the Pinot he left his life for in New York.

Every minor change in his winemaking had an impact, and the cumulative effect was significant; Jamie's wines were starting to express their individual vineyard origins with the sense of clarity and definition he had been aiming for all along. To this day, the tweaking continues, and every vintage, he sets aside a few barrels to experiment with, always looking for the next unexpected discovery that will lead to even better, more subtle, and age-worthy wines.

His tenaciousness has paid off brilliantly. In recent years, Kutch Pinot Noirs have become benchmarks among Pinot obsessives.

Collectors and sommeliers regard Jamie's wines as particularly elegant evocations of Burgundian restraint that still remain true to their sunny California origins. Critics, too, have realized how important Jamie's work has become in the wider world of American Pinot Noir. Antonio Galloni, a highly influential critic who made his name with Robert Parker's *The Wine Advocate* and now helms his own deeply respected publication, *Vinous*, called the Kutch 2017 McDougall Ranch Pinot "a total stunner."[2] Jancis Robinson, arguably the most respected wine writer in the world—she was even awarded an OBE by Queen Elizabeth II for her work—praised the same wine as "quite outstanding!"[3]

Jamie now produces ten different wines, including two Chardonnays and eight Pinots. He leases parcels in a wide variety of vineyards because his goal is for his wines to provide a sensory map of sorts to the individual places where they're grown. In the Sonoma Coast, that means capturing the essence of Falstaff Vineyard's sandy soils and low elevation as well as the fog and cold that chill the clusters of grapes there. McDougall Ranch is planted at a higher elevation, in rocky soils and on a warmer site, which results in more effusively spiced fruit that feels heavier on the tongue. For Jamie, that range of raw materials—of grapes grown in such unique soils and microclimates—plays directly into his overarching goal. It's backbreaking and bank-breaking work. He spends hours each day at the winery, and just as many driving back and forth to the vineyards themselves, muddying his boots, conferring with vineyard managers, and popping grapes from their bunches to assess their progression. As the 2017 harvest approached, Jamie was feeling optimistic about the quality of the fruit on the vines. This, he thought, could be a banner year.

After rushing away from the hotel that evening in October 2017, the Kutches made it all the way back to Sonoma but couldn't get past the police and fire barricades to actually access the winery. Instead, they had to rely on Twitter updates from the few journalists who had managed to slip into the fire zone, parsing clues in their reporting to see how much they should worry about the approaching flames. It was clear by midnight that the fire was uncomfortably close to his winery on East Eighth Street, but he couldn't get any closer to see for himself.

To understand how a fire like this one exploded so quickly, we have to go back to the previous winter. Incredibly, the violent nature of what came to be known as the Tubbs Fire stretched back to early in the growing season, when severe rain drenched Sonoma County. That excessive moisture throughout late winter and early spring created an unusually dense amount of cover growth. The rains were followed by several vicious heat waves and droughts that effectively baked the grasses and wild growth like wheat in a kiln. Any spark could easily explode into flame. This is exactly what happened on a ten-acre property off of Bennett Road in Napa Valley's Calistoga, when an old utility pole that was scheduled to be replaced the following spring gave way. It had been "woodpeckered so damn bad,"[4] the caretaker of the land explained to investigators, but no one expected it to fail as catastrophically as it did.

On the evening of October 8th, gusty winds roaring in from the northeast shoved the leading edge of the fire over twelve miles in just a few hours. By one o'clock in the morning of Monday, October 9th, it had breached the city limits of Santa Rosa. Around 1:30 a.m., emergency evacuations began in and around town. By 2:00 p.m., the conflagration, which had been fanned by increasingly powerful winds, jumped Highway 101. It kept on consuming the fuel that had been created for it over the course of the brutally hot and dry growing season.

It only got worse from there. By midday Monday, the unquenchable appetite of the fire had become clear; over a thousand homes and businesses were smoldering or hollowed-out husks. Sutter Health Hospital and Kaiser Permanente, both in Santa Rosa, had to be evacuated. Some doctors actually raced away with patients in their own cars. The Oakmont of Villa Capri senior center was gone. The Vista Campus of Santa Rosa Community Health was torched. By Thursday, the fire had consumed nearly 25,000 acres and was still only 10 percent contained.[5]

Jamie finally made it past the police and firefighters and got into his winery on East Eighth Street. The initial scene was grim; he couldn't see from one side of the space to the other because of the thickness of the smoke. His winery is tucked into what looks like a massive garage, albeit one that's as large as two or three houses. Still, it's not the winery of popular imagination, with beautiful moldings and wrought-iron candelabras. Instead, it's the embodiment of what is often referred to in the business as a working winery. Steel tanks stand sentinel along the walls, oak barrels are stacked in front of them on unadorned metal racks, hoses run in a tangled spaghetti-like pile along the floor, and the air is punctuated by the regular eep-eep-eep of forklifts. Yet the wine he crafts there is spectacular, as transporting an evocation of the Burgundian ideals of restraint and elegance as any in California. Jamie's focus on the unglamorous nuts and bolts of winemaking has always been infinitely more important to him than any sense of artifice.

He grabbed his wine thief, the device used to pull wine from a barrel through the bunghole at the top. Jamie was terrified as he raised the glass to his nose for his first sniff. He knew on an intellectual level that if he had done everything right and ushered the grape juice through fermentation and into barrels before the fires swept in, the wine should be fine. But he suddenly wondered whether barrels from one cooperage over another might be more prone to absorbing the off-aromas of the fire. In the moments before he took his first sniff and sip of the wine he'd just thieved from the barrel, his heart started pounding as he pondered the possibility that the wood had absorbed the smoke and ruined the liquid within.

Jamie took a sip. Luckily, none of his fears came to fruition: the wine was safe. Other producers got lucky too. Because of the heat spikes that marked the summer, many vineyard managers and winemakers who typically pick later in the season were forced to harvest on the early side; the extreme heat led to rapid ripening and a not-too-long growing season. As a result, a large swath of the vineyards in Sonoma and Napa had been harvested by the time the Tubbs Fire exploded to life. For all the devastation of property and loss of life that it caused, wine, in general, survived. Summing up the vintage after a tasting of wines from that terrifying year, *Wine Enthusiast* magazine was effusive. "The quality was outstanding and helped allay fears of smoke-tainted wine," the critic Virginie Boone noted. "They showed the vintage's high level of quality . . . [and] though the vintage was a tricky one to manage, it was still anticipated to be of excellent overall quality before the fires added an unexpected question mark to the final equation."[6]

Of course, the relatively small number of producers who didn't harvest before the fires tell a different story. Once a massive fire erupts, it tends to be increasingly difficult to find vineyard workers to actually pick the grapes; either they struggle to get from their homes to the vineyards or the smoke is too thick to allow them to safely work. In general, however, the heat waves and drought, though also responsible for the fires themselves, ironically, saved the vintage by forcing an earlier harvest. It was a Pyrrhic victory, and one that bodes terribly for the future. Luckily, though, more wineries than expected were able to survive and fight for another year.

The Tubbs Fire was finally extinguished on October 31st, at which point it had burned nearly 37,000 acres, completely destroyed thousands of homes, and damaged hundreds more. Twenty-two people perished.[7] It was the most destructive wildfire that California had ever experienced . . . until the following year, when the notorious Camp Fire was even worse. In the final accounting of 2017, California fires burned more than 1.5 million acres.[8]

Terroir is an impossible-to-directly-translate French word that encompasses all of the external influences that impact the growth and development of grapes in the vineyard. It's generally agreed to include soil type, underlying geology, microclimate, valley floor or hillside location, drainage, the vineyard's angle toward or away from the sun at various times of the day, and more. The local winemaking culture is often also included in definitions of terroir, which is built on the idea that the place where grapevines grow is a key to how their character is ultimately expressed in the glass. Jamie Kutch's Pinot Noir from the McDougall Ranch, which is based on oceanic soils whose first deposits stretch back 150 million years, is powerful and structured by crunchy textured tannins that somehow speak to those ancient origins. By contrast, his Pinot Noir from the Falstaff Vineyard, which is planted in far younger marine quartz sandstone soils that are only around 5 million years old, is softer in texture and better suited for enjoying at a younger stage of its evolution in the bottle. The Sonoma Coast used to be part of an ancient seabed, and millions of years of geological churn have resulted in its various hillsides and tucked-away valleys. Each one, it seems, is perfect for the cultivation of remarkably different styles of Pinot Noir.

Our planet's history is one of billions of years of change, seismic activity, explosive volcanoes, and tectonic plates moving entire continents. What makes this current time so unique and frightening is the accelerating speed with which environmental and atmospheric shifts are occurring as a result of human behavior. Yet Jamie and Kristen remain cautiously optimistic. They even recently purchased twelve acres in the Sonoma Coast appellation, where they're gutting the old house on the property and preparing the land for their first vines to be planted. I spent an afternoon there with the Kutches in July 2021, and watching Clayton drive the little Kubota tractor as Jamie tossed into the back of it wood from the trees he had spent the previous week removing made me think about all of the generations of Americans who came before them, who took a chance, found land out West, and made a go of building something lasting for their families. Jamie and Kristen's bet is that building a winery there, with a tasting room and light bites for their guests, would allow them to benefit from the added hospitality revenue it would bring in. The additional income stream would also help them ride out any future catastrophic vintages.

"But it got us thinking," Jamie told me. "If the climate is definitely heading in a hotter direction, then maybe the smart move is to start looking for land somewhere that's really at the edge of where grapes can grow and produce good wine, with the understanding that in the coming years and decades, it'll actually become one of those sweet spots." There's a reason that wine pioneers have been pushing to further and further climatic extremes in an effort to get ahead of the curve. Fine wine from Northern Europe, for example, is a small but intriguing category. More and more vineyards are being planted farther south in Patagonia. Producers in both are hoping that a warming planet will make the far-off reaches of Scandinavia and South America the next big things in the world of wine.

Jamie's other option is to give up on his beloved Pinot Noir altogether and start learning how to work with grape varieties that are less finicky and thrive better under severe heat stress. Grenache or Syrah are two options if he chooses to go that route. He may live and work thousands of miles from New York City, but the Wall Street trader in him is always making calculations. Jamie is remaining realistic about what he might have to do to survive in the wine business for another twenty-five years. He's also painfully cognizant of the fact that if massive fires keep on exploding year after year, it doesn't matter what he plants or where he carves out a vineyard. A big enough fire, moving quickly and at exactly the wrong time of the season, will destroy a vintage and even an entire winery.

For now, Jamie seems to have found a way to make his current situation work. Still, the annual fear of fires is something that he realizes isn't going away. So he's making the most of it, riding his well-earned wave of popularity and doing everything he can to keep on producing the kind of wines that collectors are increasingly willing to seek out.

"I have to work hard," he told me. "I put in a large amount of labor. I don't have the money for an oenologist, for a huge team. But with that comes the word *happiness*, and I'm happy. And that trumps bringing in hundreds of thousands of dollars or selling for millions one day. . . . I've really watched others in my space, in this industry, and I see their success or failures, and I try to turn that into a question of what would be success for me." In the end, he said, "I wanted to blaze my own trail, find my own voice."

He is clearly doing something right, even in this age of risky grape growing and winemaking. For how many more years or decades he'll be able to continue operating the way he has, however, is the question. He got away relatively unscathed in 2017, but the fires of 2020 reduced his production by more than 80 percent. Given the growing frequency and intensity of California's wildfires and a climate with an unpredictable tendency to change faster than modern humans have ever seen, the future is murky.

"Nature, no matter what we do as a work-around, no matter how clever we think we are, always wins," he said. "Eventually, in a century or two, or not even that long, there might not even be a California wine country anymore."

2

WAKING UP TO CLIMATE CHANGE

Nicolas Seillan couldn't sleep. The château where he'd been living for only a few years was always drafty—that's the nature of these old, sprawling homes; they may give the impression of impermeability and permanence, which is accurate in a lot of ways, but wind and noise somehow manage to find their way through the corners with remarkable ease. He could tell that something just wasn't right. After lying in bed for a few minutes, listening to the rain thrumming outside, he got up, took a drink of water from the bathroom faucet, and went back to his room, concerned that the conditions outside could lead to hail. Before climbing back into bed, he opened the window, stuck his arm outside, and knew right away that this was not a normal mid-May storm.

Within just a few seconds, the cold rain turned to large hail, pelting his arm, drumming on the roof, pounding the stones of the walkway outside. But it was still the middle of the night and totally black out—Saint-Émilion is the kind of place where every last constellation seems to be visible in the otherwise inky sky—and he had to wait until sunrise, still hours away, to be able to assess the damage from what seemed like a serious onslaught.

With dawn came the realization that they were in trouble. "I immediately called my father," Nicolas told me. "I said, 'We have a big problem. We have a *big* problem.' It was truly devastating, and I didn't know what to do." The vines looked like they'd been through a war overnight; leaves were torn and battered, branches bent at unnatural angles, the tiny buds that would one day have become Merlot, a key part of Château Lassègue's signature red wine, bruised, some beyond recognition. That time of year, grapevines across Bordeaux are generally in a deeply energetic growth mode . . . and now they had been stopped in their tracks, like a team of Olympic sprinters who had been tackled by the defensive line of the Chicago Bears. Pierre Seillan, his father and co-owner of the estate, told Nicolas to stay calm and send him photos of the damage—he'd been through enough challenging vintages in his long career that he was confident they'd be okay. "We are going to find a solution," he told his son. But it was hard to see what that might be or how it would even be possible; the vineyard looked like some sort of vinous version of Omaha Beach on D-Day plus two.

On top of all this, 2009 was Nicolas's first year as vineyard manager at his family's estate. "It was my fire christening to manage a vineyard," he told me—his baptism by fire. "I was only considering the glamour part" of running the estate, he went on, though his past experience both there and at others in Bordeaux and the United States had prepared him for many of the challenges ahead. Still, things tend to look different when you're at the helm. "I was shocked with reality in a very harmful way." He paused. "Today, I say it with a little lightness or humor, but I can tell you, at the time it was quite alarming because you are facing a situation where you seek help, particularly when you don't have much experience. . . . It felt like I would be your captain on a boat, and you have a big hole. The water is coming in and you ask for help because it's extremely serious."

The sun was climbing by this point, and the vineyard workers had begun to arrive. As they did, the looks on their faces registered the same alarm that Nicolas was feeling. Their future, too, had suddenly been thrown into doubt.

Pierre Seillan is a vigneron, and he has the sturdy, meaty hands to prove it. The word, which really doesn't have an exact equivalent in English, loosely translates to something like "winegrower" and refers to the almost sacred concept in France that great wine comes from the vineyard, from the land itself, and not from any sort of trickery in the winery. The process of winemaking ushers the juice from the vineyard to the bottle, of course, and the best winemakers are absolutely central to the work of finding the best in a particular selection of grapes, a particular part of a vineyard, and coaxing it to the fore in the final blend, but the figurative roots of the liquid are deeply tied to the land in which the vines' actual roots are sunk. It all goes back to the quintessentially French concept of *terroir*, of the inimitable character of the place that leaves its stamp on every sniff and sip of the wine that's produced from the grapes that grow there. Terroir comprises everything from the composition of the soil and its ability to drain or retain water, to the angle of the land, the way the wind blows through the vines as a result of a particular kind of microclimate, and so much more. The job of the vigneron, then, is to usher the grapes to perfect balance and maturity (often with the help of a vineyard manager and, always, a team of dedicated workers who carry out the pruning, the vine training, the leaf thinning, and more). The vigneron helps each vine find a sense of equilibrium between ripeness and what's often called terroir specificity—overripe grapes tend to have assertive, often sweet flavors that can cover up that sense of place, whereas underripe grapes can be hard and overly acidic or astringent—and then translates it all into a wine that's reflective of both where it was grown and the character of the vintage. In that regard, a vigneron is a winemaker but also a farmer of vines and a translator of geology. There's something both earthy and mystical to the work of the vigneron, and Pierre Seillan embodies both brilliantly.

He was born in Gascony, a little more than a hundred miles from Saint-Émilion. Growing up, his family owned the small vineyard and winery of Bellevue, and his love of wine was nurtured by a father and mother who made their living and fed their family as modest grape farmers and winemakers. The grand châteaux of Bordeaux and the great houses of Champagne may get most of the attention, but historically, the vast majority of grape farmers and winemakers in France have been small-scale operators whose fortunes were so tied to the vagaries of the vintage as to be almost inextricably linked to them; the weather, the environment, it was all *personal*. Over the course of his long career (he has actively participated in or helmed more than fifty vintages), Pierre has gotten grapes to the bottle in Gascony and the Loire Valley, and then, in a professional partnership with Jess Jackson and his wife Barbara Banke (which has shaped not just the second half of his storied career but also his family's fortunes), Bordeaux, Tuscany, and Sonoma County (he had worked in Bordeaux before the partnership but not at Château Lassègue).

In San Gusmè, a sun-dappled and sleepy hamlet in Chianti, he serves as the vigneron of Tenuta di Arceno, and those wines have quietly, over the years, become the kind of reds that collectors love. In Sonoma, alongside his daughter, the winemaker Hélène Seillan, he is responsible for the wines of Anakota and the three reds produced by Vérité, which have become critical darlings—three different bottlings that each channels a unique aspect of the classic Bordeaux style through the lens of sunny California: La Muse, which is based on Merlot; La Joie, whose footing is found in Cabernet Sauvignon; and Le Désir, which is all about Cabernet Franc. They remain, to this day, some of the most highly awarded and critically acclaimed reds in Sonoma history, with a total of sixteen 100-point scores between them from Robert Parker's *Wine Advocate* over more than twenty vintages.

When he received Nicolas's call on the morning of May 11th, he drew on a lifetime of experience to understand how to handle the situation and how to move forward. Pierre could hear the fear in his son's voice. He had felt that uncertainty himself over the years and knew not just what to do but more importantly, what *not* to do. "Stay calm," he told his son. "We are going to find a solution."

The vineyard workers saw how dire the situation was as soon as they arrived, and as more and more pulled into the lot, they began asking Nicolas the obvious question: Would he be laying them off? Would he scrap the vintage, cut his losses, and focus on saving the vines to hopefully be able to make wine next year?

"You have to consider that you're paying salaries," Nicolas explained. "You have families that you are responsible for. That morning of May 11th, I was looking at them, and of course you don't need to show any sign of weakness because they were already in shock. . . . I said, 'Of course not. Don't worry. We have some solutions, and we are strong.'" But in reality, Nicolas wasn't sure what the right move was; the devastation was massive.

By midmorning, the consultants he'd reached out to had begun to arrive, and one after another told him that 2009 was done, that the best he could do was to prune back the damaged vines so that they'd be able to reserve their energy to heal, and hopefully, with luck and tenacious work in the vineyard, they would have a solid harvest the following year. But this year, the consultants stressed, was lost. (Consultants, incidentally, have become an important part of the wine world, and producers from Napa and France to Australia's Barossa Valley and beyond often work with consultants to help them dial in their work in both the vineyard and winery. The best of them are often huge helps to the estates that hire them.) Yet with decades of experience, Pierre knew better than that. He asked Nicolas to send him more photos now that the sun was arcing its way up. The morning of May 11th was cruelly beautiful in Saint-Émilion, one of those shimmering days that seems to have been sketched by an illustrator of children's books. The damage, Pierre saw, was extensive: the vines looked *hurt*. But on closer inspection, he noticed something that the consultants hadn't: a path to making it through the calamity. "Nicolas," Pierre told him, "don't prune now in the middle of May. Don't pull back." If they did that now, if they cashed in their proverbial chips and then had a bad frost in October or November, the vineyard could die. "You will kill your vineyard." All the advice that the consultants had offered, Pierre told his son—to cut off the damaged parts of each vine entirely—"was stupidity." Instead, he suggested carefully pruning each vine shoot by shoot, based on exactly where it was damaged. That would allow each one to recover more quickly and to channel whatever energy it could to the buds that had survived the onslaught rather than to healing its damaged parts. This kind of precision pruning, as opposed to effectively amputating the entire damaged limb, was like cauterizing a wound, painful and not particularly pretty, but it was their best chance of saving each vine, treating every one of them like an individual patient, and giving it what it needed to be able to recover and eventually become whole again.

This was the vineyard equivalent of battlefield medicine.

Château Lassègue's vineyards are generally so reliably washed with sunshine that the facade of the winery itself is home to two massive sundials. Bottles of their flagship *grand vin* are clothed in a label with a stylized image of a sundial, and their so-called second wine, a less expensive yet still excellent expression of their land that's composed of juice from mostly older parcels on the foothills of the estate as well as some younger vines that haven't yet reached peak levels of expressiveness, is called Les Cadrans, "the sundials" in French.

That sunshine plays a huge roll in the character of Château Lassègue's wines. Its vineyards, all sixty acres of them for Lassègue and a total of eighty-nine acres including the plots for Les Cadrans, start at the top of the hill just outside the château and extend all the way to the bottom of the slope, like an apron, with exposures that run the gamut from southeast to southwest. This gives the vines impeccable access to sunshine throughout the morning and into the late afternoon. For that reason, Lassègue has always been planted with a good deal of Cabernet Franc and Cabernet Sauvignon. Merlot is king in this part of Bordeaux—Saint-Émilion and the neighboring appellation of Pomerol have a famously high amount of clay in the soil, which is the kind of land in which Merlot reaches its peak of generosity and longevity—but in Saint-Émilion, in particular, there is also a good deal of limestone in the topsoil and underlying geological layers, which is exactly the kind of terroir in which Cabernets Franc and Sauvignon thrive. The vineyards at Lassègue, with the warmth of their southern exposures and high proportion of limestone, have always nurtured these two varieties with particular aplomb.

In historically normal years, Lassègue's grapes have tended to achieve excellent ripeness, resulting in wines that despite their age worthiness, are typically not just approachable in their youth but also deliciously generous. (Often, wines that are meant to age for a significant length of time tend to be less pleasant in their youth and take several years for the structure-giving tannins to lose their youthful tenacity and allow the fruit of the wine to shine through; producers like Château Lassègue and others have been able to craft wines that are both delicious early on *and* that often have the ability to mature for decades.) Top critics have described the best vintages of Lassègue as "opulent," "sumptuous," and "powerful," and it was for this potential that Pierre and his wife, Monique, purchased the property in 2003 after an extensive search with their partners Jess Jackson and Barbara Banke. They walked the land of vineyard after vineyard, searching for that inimitable *thing* that would set theirs apart. Once they saw Château Lassègue and studied the land surrounding it, they knew. "It's that estate and not another one," Pierre proclaimed. (Ironically, 2003 turned out to be a vintage marked by a heat wave that killed thousands in France; challenged the skills of vignerons throughout the country; and in Bordeaux, as elsewhere, resulted in wines of marked power and often, flamboyance—a controversial style that some critics derided as not like Bordeaux at all but rather more akin to Napa Valley. In the world of wine, that amounts to a gut punch, claiming that wines from such a historically important region don't taste like themselves. This question of climate change's impact on the character of the final product in the glass is one that comes up with increasing frequency and has for some time now.)

But back when the Seillans, Jackson, and Banke were considering purchasing Lassègue, Pierre had no way of knowing what 2003 would bring, much less 2009. What he *did* know was that Lassègue had too much going for it to pass it up; he was particularly taken by a few blocks of old vines that had been planted before the notorious frosts of 1956. Those horrifically timed cold snaps destroyed vines throughout Bordeaux that year, and their effects were still being felt on the Right Bank, where Saint-Émilion is situated, several years later. Amazingly, in the wake of the hailstorm of May 11th, those ancient vines would be what taught Pierre, Monique, Nicolas, and the rest of the family what they had to do moving forward. The past, at Lassègue and throughout so much of the wine world, would end up pointing the way to the future.

"There is a very natural way to assess the impact," explained Nicolas Seillan. "You need to let the sun go down on your vineyard." He wasn't speaking metaphorically, although given the violence of the hailstorm the night before, sacrificing that vintage—letting the proverbial sun set on it—had been a real possibility. What he meant, rather, was that it was imperative to give the vines a few days to be accurately assessed, to respond to their injuries, and to watch them throughout that time, following "the change of color of your vineyard [so] you can really see what is still [viable] or not," he elaborated. After the violence of the night before, the vines were "in shock," as Nicolas put it. Some had their canopy of leaves, which are the engines of photosynthesis and what essentially power the production of sugar and fruit, completely broken—torn to shreds, in some cases. Others, however, looked to have been more damaged than they actually were. The day after a weather event like this, grapevines react in much the same way as a human being does to a traumatic injury, and the initial impression may not be the right one; you have to wait, in some cases, for the proverbial swelling to go down. When it comes to damaged vines, you have to wait a day or two to get a more thorough read on the situation.

But it's also important to act fast because if you do nothing and the injuries are in fact that threatening, it's imperative to start helping them as soon as possible. In the case of vineyards, that can mean spraying them with a copper solution to help stave off any mold or fungus that could potentially get into the vines themselves, which would leave them open to a secondary trauma—infection.

"We were really only able to assess the issue the next day," Nicolas recalled. "I could see which blocks were particularly affected, and I could make some kind of a report to my father. . . . And luckily, I was able to realize that some important blocks were still okay." This, incidentally, was not the case for everyone—like so many catastrophic weather events, the damage seemed to be random, like reports after a tornado in the Kansas countryside when bewildered neighbors marvel at the capriciousness of one house being demolished and the one next to it remaining completely untouched. In Saint-Émilion, some vineyard blocks looked like they'd been through an aerial bombardment, and others remained unscathed.

As the days wore on, though, the damage to certain parts of Lassègue turned out to be as serious as it initially seemed. "Only after a week," Nicolas said, "[did] we realize that the vine was still in shock. . . . The vineyard was not growing. The vine was blocked." The consultants again recommended that the Seillans cash in their chips, do a severe pruning to save the vines, and hope for a better year in 2010. "'You have only one solution,'" Nicolas recalled one of them informing him with all the solemnity of a doctor talking to a patient's family after an exam that turned up a particularly grievous condition. "'You re-prune everything completely. This year you won't have any crop, of course. Next year, you will have only a small percentage, but you need to save your vineyard.'"

It wasn't bad advice, especially given the context; the damage that some of Lassègue's neighbors experienced required relatively drastic action. But the Seillans knew their vines, their land, and decided to ride it out for another couple of weeks. They still had hope. And indeed, nearly ten days later, Nicolas and Pierre noticed a change on their land. "It woke up again," Nicolas told me, marveling. "The vines went back to life."

Some of that has to be attributed to blind luck—that's the nature of violent storms—but Pierre is convinced that there was more at work there. Lassègue's old vines are "a great witness of the last century," he argued. Old vines, Pierre firmly believes, retain a sort of chemical memory of the ways in which Mother Nature has challenged them in the past; they somehow learn how to make it through all but the most catastrophic of weather events. They're like crusty old soldiers in that regard, less flappable than their more youthful counterparts. Old vines also have deeper, more complex root systems, which allow them to pull from a literally and figuratively deeper well of water and nourishment to get what they need to survive. "Those old vines have the capacity not to fight—because you don't fight a hailstorm. You just take what falls on you—[but] the vines were able to find the energy in themselves," Nicolas explained. "If we had five-year-old vineyards, ten-year-old vineyards, it would have been another consideration," and they likely would have been lost, either for that vintage in particular or perhaps forever, requiring replanting . . . at great expense. But the high percentage of older vines on the Lassègue property was a hedge against a situation like this. They also realized that as they were able to assess the damage at a more granular level, block by block and then vine by vine, the ones that fared best were Cabernet

Franc and Cabernet Sauvignon, which changed the character of their wines afterward; it showed them how their plantings, and therefore the ultimate blend of the *grand vin* and Les Cadrans, would have to change as the climate became more and more unpredictable.

Bordeaux, like virtually all wine regions around the world, has a long history of dealing with challenging weather. (I say "virtually all" because until fairly recently, Napa and Sonoma, for example, benefited from almost oddly propitious climate patterns. There were bad years, of course, but the run of so-called good ones far outnumbered them. Recently, more common and vicious droughts, heat waves, and wildfires have changed the calculus entirely.) Bordeaux is situated in southwestern France, a little over three hundred miles from Paris in the Gironde estuary, and its vineyard land stretches over 280,000 acres. The region is parceled into fifty-seven officially recognized and regulated appellations, though its most famous division is far simpler than that; the Left Bank, as it's known, is home to some of the finest Cabernet Sauvignon–based reds in the world, and names like Château Lafite Rothschild, Château Margaux, Château Latour, and more dot the landscape of the Médoc, the most prestigious part of the Left Bank. The Right Bank, on the other hand, is where Merlot and Cabernet Franc excel. There, the greats of the Pomerol appellation, like Château Pétrus, not only represent the heights that Merlot can achieve but also command some of the highest prices of any wines on the planet. A bottle of 2018 Pétrus can be found for $4,500 or more, and older vintages can easily double or triple that. The neighboring appellation of Saint-Émilion, where Château Lassègue is joined by Château Angelus, Château Cheval Blanc, and other highly regarded names, is Cabernet Franc country (though Merlot also does very well there too). The geology is different on this side of the river, and the soils allow Cab Franc and Merlot to shine with a particular sense of expressiveness and complexity. The wines from the Left and Right Bank couldn't be more different, and similar dramatic differences can also be found with the reds from Pomerol and Saint-Émilion. These bifurcations mainly come down to the soil, to the underlying geology of the land that dictates which grape varieties will thrive. The gravel (and sand, clay, and limestone) of much of the Left Bank is perfect for Cabernet Sauvignon—that gravel, in fact, reaches its peak in the appellation of Graves, which means "gravel" in French, and is where the legendary Château Haut-Brion is grown, in the part of Graves known as Pessac-Léognan. The clay and limestone of Saint-Émilion provides a perfect environment for Cabernet Franc vines to sink their roots into, and the clay of Pomerol is perfect for Merlot. Pétrus, in fact, is grown on a vineyard that boasts what is called a "buttonhole" of clay that rises up from the underlying soil, lending the vines there a thoroughly unique character. Walking the vines at Pétrus, in fact, left my shoes tinted with a layer of fine, clay-based dust.

Yet beyond the great appellations of the Right and Left Banks, there is an entire extended world of Bordeaux—the region, as I've said, is huge, and it can take hours to drive from one end of it to the other. Entre-Deux-Mers, between the Garonne and Dordogne, produces wildly underappreciated white wines that often represent some of the best values in the region and some of the most delicious expressions of Sauvignon Blanc and Sémillon in the world. More southerly appellations, like Sauternes, Barsac, and Cadillac, are home to ambrosially sweet wines that leverage the fog and humidity that settle over the vines in the autumn to promote the growth of the *Botrytis cinerea* fungus on the Sauvignon Blanc, Sémillon, and Muscadelle grapes, shriveling them up, concentrating their sugars, and lending the wines that result from them distinct notes of honey, sweet spices, and a pulse of earthiness. The vast swath of land north of the Right Bank, where wines simply labeled Bordeaux or Bordeaux Supérieur are often grown, is where consumers can experience the telltale character of Bordeaux without spending a fortune; $25 can buy an excellent bottle of red Bordeaux that vastly overdelivers for the money.

This divvying up, this parceling of the land in Bordeaux, is typical of not just the world of French wine but also of wine regions around the world. The process of discovering which hillsides, which soils, which bends in the river will be most suited to one type of grape, one style of viticulture over another, is one of the keys to success. The nature of fine wine—and by this I don't mean expensive, necessarily, but rather wines that are intimately tied to the land in which they're grown, as opposed to bulk produced from grapes grown over a wide stretch of a region or entire country and brought together in a factory-like winery—is based on the idea that major and more minute differences from one appellation to another, one vineyard to another, sometimes from one row of vines to another, have a dramatic impact on the wine that is ultimately produced from the fruit that's grown there. In fact, even the term that's employed for people like Pierre Seillan, *vigneron*, references this concept that great wines aren't so much made as grown. Caring for the vines and crafting wines from their juice cannot be separated from one another.

Yet for all its elevated reputation as a region, and for all of the often stratospheric prices that the best wines of Bordeaux command, it is still a maritime-influenced climate that has historically fallen victim to the whims of nature. Even the fact that the wines of Bordeaux are almost always blends, as opposed to single grape varieties, is proof of the fickle nature of the climate there. There are six main permitted red grape varieties in Bordeaux—Cabernet Sauvignon, Cabernet Franc, Merlot, Petit Verdot, Malbec, and Carménère—which act as insurance policies for one another. Each variety flowers, buds, develops, and ripens at a slightly different time and pace, which means that if, say, the Merlot has already started to bud and is irretrievably damaged by hail, a vigneron may still have Cab Franc, for example, to lean on for that vintage since it tends to develop a week or two behind Merlot. Now, in light of climate change, a handful of other grape varieties are going to be permitted in wines labeled as Bordeaux AOC or Bordeaux Supérieur, though they can't comprise more than 5 percent of the blend. Still, it's telling that the authorities have called Arinarnoa, Castets, Marselan, and Touriga Nacional—all warm-climate varieties—as grapes that could play a greater role in the future. (Interestingly, Marselan is also being discussed in Israel as one of the "grapes of the future" there in light of the heating-up and weather-weirding world.) The great vineyards of Bordeaux, then, are most easily understood as being planted to specific percentages of each of their constituent grape varieties, but the actual blends of the wines shift every year depending on how each variety did during that particular growing season, the specific characteristics they'll bring to the blending table, and the ways in which the winemaker feels he or she can best express the character of the land as seen through the lens of that vintage.

In fact, the expression of any given year is tattooed on the wines in ways that are readily apparent if you know what to look for, and tasting a lineup of great reds from Bordeaux is a lesson not just in the pleasure and range of what the region is capable of but also in the impact of weather and a changing climate over the decades. Bordeaux, located where it is, swept by storms and sunshine in often dramatic measure, has always been at the mercy of the elements. Early-spring frosts can diminish a crop substantially; hail can do the same and also cause enough damage to the vines themselves that vignerons have to worry about the following year's crop; and heat waves can overripen fruit, force early harvests, and shut down the vine's ability to keep on directing energy to the grapes. Torrential rains just before harvest have the potential to plump the berries with water and dilute their flavors and aromas. Bordeaux, in other words, is no stranger to so-called bad years. But the good ones have the potential to be truly magnificent, producing wines that can age for decades with ease and develop unforgettable flavors and aromas. What makes the current climate situation so challenging is the fact that the extremes are becoming even more so . . . and occurring with ever greater frequency. Fortunately, the overall quality of the wines seems to be improving . . . for now. *Wine Spectator* rated all but two vintages of Left Bank Bordeaux ninety or higher on the 100-point scale between 2009 and 2020, and only one vintage below ninety points on the Right Bank during that time. One of the advantages of a warmer climate is riper fruit, more generous textures, and the kind of tannins that barring excessive heat, can facilitate graceful and long aging while still allowing the wines to be approachable in their relative youth. (There is, however, a great deal of discussion about this, and many producers believe that slightly cooler vintages can age for even longer, despite generating less buzz *en primeur* or on release.)

But more and more often, unusual and potentially catastrophic weather events threaten producers throughout the country. The Bordeaux reds of 2003 are famously powerful and heady—they're also controversial, as I've noted, with some critics warning that they occasionally resembled reds from California (or even resembled Port) more than classic Bordeaux[1]—but the heat wave that broiled so much of Europe that summer was responsible for over fourteen thousand deaths in France and more than thirty thousand throughout Europe.[2] On both the Left and Right Banks, 2009 is now considered a great vintage, but lost in the praise of the wines is the fact that a number of producers made less than they typically do, the result of loss from the May hailstorms. In March 2021, a freak heat wave, followed by widespread frosts, led the French government to declare an "agricultural disaster." According to *Food & Wine*, "temperatures in parts of France soared to highs of 26°C (79°F), which caused some crops—including grapes in the country's famed vineyards—to start blooming early. That made the situation even more devastating when temperatures plummeted to -7°C (19°F). . . . That heavy frost affected up to 80 percent of French vineyards in almost every region," leading to fears that the "year's grape harvest could be one of the smallest ever."[3]

The article went on: "Some winemakers did what they could to try to salvage their crops, lighting candles and small fires in desperate attempts to keep frost from forming. In many cases, it wasn't enough. 'We'd bought huge candles—like big paint pots full of wax—and we placed them between the vines and ran out to light them at 2 a.m.,' Michel-Henri Rattet told *The Guardian*. 'There were still some green shoots but then the snow came. It was catastrophic.'" Images of vineyards glowing with fire pots ricocheted around the wine world, the beauty of the scenes all the more heartbreaking given the sense of long-term futility that they implied, especially considering that most winegrowers in France are small (or small to medium) in size. The reputation of French winemakers, in general, and the Bordeaux-based ones, in particular, may be dominated by the image of the gentleman farmer who, after a thoroughly civilized day of work amid the vines, retires to his ancestral château for some roasted duck and a bottle of claret, but the truth is that most—even in Bordeaux—are just a few bad vintages away from disaster. Losing vines in hailstorms, like the ones that pelted them in May 2009, or in abnormally warm spells too early in the season followed by frost, as happened in 2021, can potentially ruin a farming family of modest means, especially if they are followed by another difficult vintage.

There was another hailstorm on May 13th, but the damage was minimal at Lassègue compared to the one on the 11th, and within a couple of weeks, the vines that *had* survived at the estate were making a solid recovery. In fact, the shock that they'd had ultimately resulted in a long-term benefit for the wine, to everyone's surprise. "Several weeks later, progressively," Nicolas Seillan explained, the vines fully "woke up again. The vine went back to life, and was able to recuperate and continue." The consultants were still pushing to have them prune away that year's potential in an effort to save the 2010 vintage, but ignoring them and going with what Pierre had learned over a lifetime of internalizing the rhythms of a vine's life cycle was the best thing the family could have done.

Once those vines came back to life and started their usual work of drinking sunlight from above and water from below to grow and mature their grapes, Pierre and Nicolas noticed that they were approximately two weeks behind their usual development and ripening schedule. Everything else about their rhythm that year—the growth and development of the canopy of leaves, the flowering and budding of fruit and its turning from green to red (a process the French call *veraison*), the accumulation of sugar and diminishment of acid in each berry—progressed at its normal pace. . . . It was all just two weeks late, which might not seem like all that much, but aside from dramatic events in the lives of vineyards and the vines that are planted in them—hailstorms, for example, or sudden flooding rains or fires—everything about growing wine grapes is about accumulations of sunlight, of heat, of water. The overarching character of a growing season is determined just as much by catastrophic weather as by the small details of climate that add up in surprising ways. An added half-degree centigrade each day over the course of a couple of weeks of growing can be readily smelled and tasted in the finished wine. The result of the Seillans having been forced to harvest later in 2009 meant that the vines were exposed to more of the late-summer's heat as well as to the first few cooler days of autumn, which together led to excellent physiological ripeness and balance. After all the devastation and fear of the storm's immediate damage and its sending of the surviving vines into shock, the fruit that did get harvested was excellent.

The impact of the storm also showed them the importance of shifting their focus to Cabernet Franc and Cabernet Sauvignon, which develop a week or two later than Merlot, and are therefore often more protected from early-spring frost and hail. It also highlighted the importance of old vines and natural farming.

Agriculture, after all, is subject to the trends of the time, and growing wine grapes, perhaps more than any other form of agriculture, is built on the idea of coaxing out of the fruit a very specific set of flavors and aromas. Fermentation doesn't just change the character of the juice itself; it also serves to highlight aspects of it that may not have been evident before the yeast went to work. Add that to the fact that wine—and high-end wines in particular, especially those from historically venerated areas of the world like Bordeaux—is obsessively analyzed and parsed by an army of critics, journalists, and passionate collectors, and you have a situation in which the final character and ultimate success of a vintage is obsessed over in a way that, say, bananas or berries or pears generally aren't. If one year's wine is subpar, then it becomes that much harder (and more important) to keep consumers coming back for the following one in an increasingly crowded market. Consistently good quality is key, and for a long time, winegrowers strived for that above all else, no matter what it took to get there.

Until fairly recently, the methods of modern grape growing were aimed at exactly this outcome. Herbicides were aggressively used to clear the rows of anything but the vines themselves; pesticides were applied to kill off insects. Those iconic images of pristine rows of vines marching along valley floors and up hillsides with no plant or perceptible animal life to get in their way are actually photographs of a land divorced from its natural state as dramatically as if it had been clear-cut. And the results were, in hindsight, predictable in the long term: exhausted soils, local ecosystems thrown off balance, and vines whose roots didn't reach quite as far down as they otherwise would and could have. In one of the most extreme cases, the vineyards themselves were used as dumping grounds for the trash of a nearby city. In his influential Vinography blog, the highly respected writer Alder Yarrow wrote in a 2015 piece: "Many of Champagne's most storied vineyards are seemingly lifeless other than the vines that emerge from the soil, and to add insult to their chemically denuded injury, they are quite literally covered in trash." He goes on, "Stripped bare of any grass, weeds, or cover crop by decades of herbicide use, the soil of Champagne is littered with broken glass, broken porcelain, bits of plastic, cigarette butts and many other small inorganic objects. In some cases, the broken glass is so thick on the ground that the whole vineyard shines from afar, glossy with reflected light. Those vineyard rows that do not shine often have no luster because the broken glass has been covered by a thick layer of shredded tree bark, whose purpose, I was told, is to prevent erosion of the bare soil between rows. Without this protective cover of bark, the kind you might find in any landscape store, the soil washes away too easily in the frequent rain. Looking carefully at the rows of vineyards that lack such addition of bark, you can see this in action, as the ground dips in shallow saddles between the rows of vines."[4]

It may seem insane today, but there was a certain cynical logic to it all at the time. Paris had a lot of garbage, Champagne needed compost, and an agreement was made: "In exchange for paying only the costs of transportation, the vineyards of Champagne could have all the compost they wanted, straight out of the rapidly growing trash heaps of Paris. It was a win-win. Paris got rid of its trash, and Champagne got free fertilizer," wrote Yarrow. Early on, soon after the turn of the nineteenth to the twentieth century, it made sense on the surface, Yarrow pointed out. The kind of trash that was shipped to Champagne was fairly innocuous and likely to break down in the vineyards: food scraps, non-chemically–treated clothing that had been disposed of, that sort of thing. But eventually, Yarrow wrote, "The petrochemical revolution brought about by the Second World War introduced the rapidly modernizing world to plastics and all manner of disposable inorganic materials, all of which were ending up, broken, shredded and pulverized, in the dumpsters of Paris and therefore also in the trucks that headed periodically to the vineyards of Champagne." This Paris-to-Champagne trash transfer came to a halt in 1997, but the damage was done, and in dramatic form; even in early 2020, during a visit to Champagne not long before the pandemic, I walked some of the most prestigious vineyards of Champagne and heard the crunch of decades-old plastic underfoot. Red and blue shreds decorated some of the soil, and minuscule pieces of plastic and glass glimmered in the afternoon sun.

There were, of course, plenty of stunning wines that were being made during this period of willful ecological ignorance, but something had to give; the land wouldn't keep on providing in the same ways it always had if this went on, and consumers would eventually grow horrified to learn that the grapes for their favorite wines were potentially being grown in other people's garbage. The dawning awareness of the inner workings of our collective food systems and the growing demand that the food we eat be more responsibly grown and raised eventually found its way to the world of wine too. Today's focus on natural wine is the logical conclusion of this, but a decade or two ago, the fight was for more modest victories: grapes grown in healthy land. And the results of that fight were stunning.

Champagne may have been particularly egregious, but mistreating vineyard land was far more widespread than most wine lovers ever knew. Which is why the shift to sustainable farming and then on to the more rigorous paradigms of organics and biodynamics was so important; it not only allowed the land to begin to repair itself but also brought back the ecosystems that had been adversely impacted from decades of abuse. Healthier soils tend to erode less, and a healthy population of beneficial insects and their predators make for land in which biodiversity can thrive. When vines can sink their roots more deeply into the soil, naturally accumulating both nutrients and water over the years, they tend to recover from traumas like the May hailstorm more quickly.

As a week passed from the night of the storm in Bordeaux, and then two, Pierre and Nicolas saw firsthand the benefits of treating their land with this kind of respect. Once they had gotten over their shock, the vines began to grow again. And the fruit, as it matured, was gorgeous. The Seillans were working on minimizing chemical inputs in their land well before 2009, but in light of all that happened that year, they've redoubled their efforts to grow their grapes in as natural a manner as possible. It's better for the land, the wine, and their long-term prospects, especially as the climate gets stranger and less predictable year after year.

After the second hailstorm that May, the rest of the 2009 growing season was devoid of serious drama. Sunlight accumulated in a more or less predictable manner, rains came at the right time, and the fruit—though diminished by around 30 percent at Château

Lassègue—remained healthy. Still, it was clear that this wouldn't be a normal year.

Harvest began on October 5th, and the fruit was picked in its entirety by the 15th. It was evident throughout the sorting process, however, that things were different from past years; because of the two-week delay in ripening that resulted from the Cabernet Sauvignon and Cabernet Franc vines' pause in development following the shock of the hailstorm, the grapes had achieved a level of expressiveness that they hadn't often in the past. This, coupled with the diminished amount of Merlot, meant that the blend of their *grand vin* would be slightly changed in 2009, with Cabernet Sauvignon and Cabernet Franc playing relatively significant roles. (The final blend is 60 percent Merlot and 20 percent each for the Cabs.) In a great year, Pierre explained, it's not uncommon for Merlot to accumulate enough sugar that it achieves 16 percent alcohol or higher once crushed and fermented. "For us, we didn't have that," Pierre explained. "Our Merlot was about 13.8, 14 percent, and the Cabernet [Franc] was about 13.5 percent and Cabernet Sauvignon 13.2 percent, something like this. . . . If you open a bottle of Lassègue today," Pierre told me, "number one, we had at the time more minerality because the Cabernet provides more minerality. The Merlot provides more softness, power, and alcohol, and polyphenol." What started off as a potentially year-ending situation has resulted in a wine whose evolution is absolutely beautiful, an elegant red with a deep sense of finesse and the potential to age for another decade at least. When I tasted it again in December 2021, I was struck by its balance of freshness and concentration as well as the particularly beautiful ways in which the ripe berry and cherry fruits are balanced with more minty and savory flavors. This is a wine that has stood the test of time in ways that no member of the Seillan family could have dreamed of the morning they walked among their decimated vines.

But perhaps more important than the wine the Seillans produced at Château Lassègue in 2009, the storm, and the lessons that were learned in its aftermath, were "a confirmation of our philosophy, which is essentially is that there is no set protocol in place," Pierre Seillan explained. "It was a reminder that we need to always remain flexible and nimble, that we must always be ready to adapt to the situation as it presents itself, and ultimately learn from the experience."

In the years since, they have pulled out several blocks of Merlot on the property and replaced them with Cabernet Sauvignon and Cabernet Franc. This is both a direct response to the ways in which the increasing heat of recent years has made Merlot so ripe and full of potential alcohol that it had begun to play too assertive a role in the final wines and to the ways in which later-ripening Cabernets Franc and Sauvignon are better protected against early-springtime storms. It also is a response to the character of the land itself; since it's located at the top of the hill and has such generous south-facing exposures, it has the ability to ripen Cabernet more fully than if it had more northerly exposures. In that regard, their response to a changing climate, based on the unique, specific conditions of their land, is very much in line with what other producers around the world have had to do.

Of course, there is no paint-by-numbers solution. Mitigation of the most disastrous effects of climate change is often discussed in terms of a wide-ranging geographic area—in France do *this*, in Argentina do *that*—but that's not how it works on the ground. Policy-wise, perhaps, that's the best way to move forward for the governments involved in the big-picture decisions that affect agriculture across the breadth of the land, but farming, especially something as fickle as the growing of wine grapes from a single property, is hyper-local. The Seillans' actions have been based on the requirements of their own land; many of their neighbors have undertaken slightly different ones. The one universal, from Saint-Émilion to Sonoma to the Barossa Valley in Australia, is the understanding that the old assumptions can't always be relied upon anymore, and that what had been the status quo for so long is no longer necessarily tenable . . . or even, in stark terms, financially viable.

In the years since 2009, Château Lassègue has benefited, like other larger and more modestly sized producers in Bordeaux, from a series of great vintages . . . punctuated, of course, by less-than-stellar ones. Yet aside from 2013, which isn't generally looked at with great admiration, the past decade has been a more or less steady stream of good-to-great years on the Right Bank. But things have changed; vignerons like Pierre Seillan have had to rethink what they plant and how they manage their vines and occasionally modify their blends to make the most of the increasingly warm growing-season weather and often unpredictably cold—and shockingly timed—lows. Those fire pots glowing in the vineyards in the spring of 2021, that attempt to keep the just-flowering vines from literally freezing to death during an abnormally aggressive cold snap in April, will likely become a far more common sight in coming years than anyone wants.

That's the nature of climate change and how it affects the world of wine; its impacts are unpredictable, yet the producers who will survive to fight on into the future and continue to make great wine will be the ones who are willing (and financially able) to make the changes that Mother Nature dictates. The Seillans, in many ways, have benefited from the hard lessons of climate change but only because they recognized the problems they were facing and were willing to pivot. They also had the financial wherewithal and genuine support of a large, well-funded company behind them, which is something that many smaller producers can't rely on in times of crisis. For most everyone, however, the lessons are the same, and Pierre has been absorbing them his entire life, ever since he was a boy in Gascony learning at the feet of his father: vignerons live to fight another vintage only when they respect the power of nature, listen to what it's telling them, and make the changes that it insists upon. Now, more than ever, that's the only way to build a life in wine, to support a family and a team of employees, to be a good and responsible steward of the land, and to produce the kind of bottles that collectors are willing to spend money on, no matter what challenges may come up along the way.

As the sun rose over his family's hail-pelted vines on May 11th, Nicolas Seillan had no idea what would become of his first vintage at the helm of the property. On the 12th, he genuinely feared for its future. Yet like so much else in the world of wine, things weren't quite what they seemed, and the wines produced by his family that year were fantastic. The unpredictability and capriciousness of climate change are its very hallmarks. How it continues to specifically affect vineyards around the world over the course of the next ten or twenty or fifty years is anyone's guess, even in the regions that have been so consistently excellent for centuries.

3

DESERT ROOTS

Wine has been made in Israel and the greater Middle East for thousands of years. Wandering through the temples and tombs of Egypt, right before the pandemic made that sort of travel impossible, I was struck by the frequency with which hieroglyphic figures were portrayed with a drink in their hands. In the case of the ancient Egyptians, it was generally either wine or beer they were imbibing, depending, many scholars believe, on their social status.[1] In Israel a year and a half later, right before the Delta variant struck—clearly my travel timing to that part of the world leaves something to be desired—Amichai Lourie, the protean figure behind Shiloh Winery, excitedly led the way on an all-terrain vehicle to a grape press on his property in Shomron, the Hebrew name for Samaria. When we arrived at our destination, dusty and sweaty, I was rendered speechless: Right at the edge of one of his vineyards, a stone basin rested in the earth. It had been built more than a millennium ago, the oldest piece of winemaking equipment I had ever seen. Other ancient presses on the estate trace their origins back to biblical times. The Torah often reads like an ancient collection of *Wine Spectator* magazines, with references throughout that extol the mystical and holy properties of fermented grape juice. Even today, as I write this in the Hebrew year of 5782—or more familiarly,

2021—Jews around the world mark the milestones of life with wine: newborn boys, just before being circumcised, are given a finger dip of wine; Shabbat is rung in and ended with a glass of wine; newly minted men and women are permitted to lead the *hamotzi*, the ancient prayer over wine, at their bar and bat mitzvahs before taking a sip of their own; and the Passover seder is punctuated with four glasses throughout the meal—one of them with ten drops ritually removed, a reminder of the suffering of the Jews and Egyptians as a result of the same number of plagues brought down as punishment for the Egyptians' enslavement of the Jewish people. Ancient texts and monuments, and religious rituals that continue today, make a none-too-subtle point: wine has been integral to this place and its people for an incomprehensibly long period of time.

The Middle East is home to a tradition of grape growing and winemaking that stretches back thousands of years. In *The Wine Route of Israel*, Adam Montefiore, the world's foremost expert on the wines of Israel and author of several standard reference books on the subject, writes: "The oldest grape pips found in the regions of modern Turkey, Syria and Lebanon date back to the Stone Age period (c. 8000 B.C.E.). The art of winemaking is thought to have begun in the area between the Black Sea, the Caspian Sea and the Sea of Galilee. Indeed, the oldest pips of 'cultivated' vines, dating to c. 6000 B.C.E., were found in Georgia. . . . The biblical Noah was the first recorded viticulturist who, after the flood, 'became a husbandman and planted a vineyard.' . . . In about 1800 B.C.E. there was a communication which reported that Palestine was 'blessed with figs and with vineyards producing wine in greater quantity than water.'"[2]

Even kosher law, as it pertains to wine production, developed in such a way that it is both beneficial to the land in which the vines are planted *and* to the final quality of the liquid itself. The *shmitta*, or sabbatical year, gives the earth a chance to recover one out of every seven vintages. The kosher prohibition against producing wine from vines before their fourth annual crop is accepted wisdom throughout most of the wine world today. Cleanliness in the winery (and so much of kashrut law is rooted in cleanliness, at least as it pertained to the ancient understanding of the world) has historically ensured a better, more consistent product. The great minds of the past often turned their thoughts to wine and to agriculture and how to improve them both. Kosher wine law is evidence even today of what happens when those efforts succeed.

Today, of course, the wines of Israel suffer from a bit of a problem of perception; they're often sold in the "kosher" section of wine shops in the United States alongside the syrupy sweet and confected bottlings that are typically not made in Israel at all—Manischewitz, perhaps the most famous example of kosher wine, is produced on the East Coast, thousands of miles from Israel. (It's also about as representative of great kosher wine as, say, an old Twinkie is of the work of a trained French pastry chef.) But the true wines of Israel, the high-quality ones that are rooted in conscientiously planted vineyards and lovingly tended vines, have seen a resurgence in recent decades. When I visited the country on a wine-tasting trip in 2012, the overall quality of the wines was relatively high on a technical level, though the majority of them had a sort of amorphous character, tying them not as much to that particular land itself as to an international concept of ripeness and oak-based sweet spice. By my most recent visit, in the summer of 2021, an inflection point had clearly been reached, and every single day I was there, I savored reds and whites of idiosyncratic character, great potential longevity, and the kind of layered complexity that is the hallmark of top wine-producing countries around the world.

It may have taken a few thousand years, but this spit of land sandwiched between the Mediterranean Sea and the Jordan River had finally arrived again on the world wine stage. Unfortunately, climate change is threatening it all in ways that invaders and malevolent leaders never could have even approached in the past.

In the popular imagination, the Middle East is a vast desert region, largely incapable of growing much aside from strife and extremism. But it's a far more complex part of the world than that; from Egypt to Syria and Israel's Mediterranean coast to the eastern reaches of Jordan, there are huge differences in geology, geography, culture, cuisine, and more. Israel, as a historic crossroad of so many of the forces that have shaped that part of the world for thousands of years, distills a great swath of it, compacting into one tiny country many of the landscapes and contradictions and foods and conflicts that define the Middle East.

The country is approximately the size of New Jersey, yet within its 290 miles of land north to south, and just 85 miles across at its widest point, the differences in landscape and climate are significant. Up north, toward the Israeli border with Syria, the Golan Heights reach their peak atop Mount Hermon at more than 9,200 feet above sea level. The climate there is significantly shaped by that altitude, and the forests dotting the landscape, combined with cold nights that in the winter can plunge below freezing, make it one of the most exciting wine regions in the country. Compare that to the Negev wine region, named for the desert that spreads out across 50 percent of Israel's landmass; this is a place of sandy soils and hot daytime temperatures as well a problem unique in the world of wine, according to Wine of Israel, the national trade group—camels that snack on the leaves of grape vines. But even here, in this more classically desertlike climate, hot days and cool nights provide exactly the kind of all-important diurnal swing (the stretch between daytime highs and nighttime lows) that grape vines need to produce balanced wines, ripe with fruit flavors and energetic with acidity.

Israel, in other words, has all of the natural elements for a high-quality wine industry, which is one of the many reasons it's been such an important part of the culture there for so many thousands of years. Still, growing wine grapes in Israel isn't without its challenges, and the natural heat and pounding sunshine of the region (this is still the Middle East, after all) combined with the lack of sufficient rainfall in many parts of the country have made adaptation and creativity as crucial to the growing modern Israeli wine industry as anything else . . . perhaps even more so. Fortunately, Israel has always been a leader in agricultural technology, which is one of the keys to its continued success today and in the future. "Israeli technology is very advanced," Montefiore told me. "We have a high-tech nation, we have people that are very inventive when under pressure. Very creative. We have the best farmers in the world. We made the desert bloom, [and] we've created lot of the state-of-the-art equipment that has now become commonplace" around the world.

The country has always been like that. I remember during my first visit, in 1991, being struck by almost miraculous-seeming patches of green within vast stretches of sand, plots of tomatoes, cucumbers, and more sprouting incomprehensibly from what looked like the most inhospitable environment I'd ever seen. Israeli technology made that happen. A focus on agricultural technology is baked into the Israeli character because when the state was created, the early leaders realized that they'd have to be as self-sufficient as possible. Their situation—geographical and political—forced the issue, which the country benefits from even today.

It's also helped make Israel one of the world's leaders in the agricultural technology sector—both in its development and in using it. Drip irrigation was created here, and the use of drones to map out and understand individual rows and vines is increasingly being employed. Dendrometers measure water uptake by monitoring the wood of the vine itself. Sensors placed on various parts of the vine allow for a remarkably precise understanding of exactly what is happening inside each one, which allows grape growers to better tailor their actions in the vineyard to more specifically meet the needs of those vines, of that land. The level of knowledge afforded by the technology being employed here is beyond impressive in its granular detail and has helped position the country to be better prepared for a world of rapidly, and often violently, changing weather.

Yet technology and ingenuity can only go so far; they are not capable of replacing the natural defenses that grapevines have evolved over the millennia. Rather, they work alongside them. Which is why, around the world, top viticulturalists are increasingly coming to the realization that everything they do should be at the service of maximizing the natural inclinations, as it were, of the vine and to getting back to as natural a state as possible in the vineyard . . . especially in this era of drastic, dramatic climate change.

In 2012, Michal (pronounced *Mee-CHAL*) Akerman, now arguably Israel's top viticulturalist, the CEO of Tabor Winery, and a renowned expert on the integration of winegrowing into the natural environment, found herself in conversation with a few members of the Society for the Protection of Nature in Israel (SPNI). They were walking a vineyard at Tabor Winery, one she had been farming in the same way that generations before her had; the rows of vines were neat and straight, with nothing between them but soil devoid of grasses, weeds, and flowers. It was a postcard-perfect image of what vineyards, at the time, were "supposed" to look like. Little did she know that that walk with members of SPNI would change her life—and totally unexpectedly, the world of Israeli wine in general.

"One of the agronomists there, he knew me, and he came to me [and] planted this idea in my mind, but he did it very beautifully," Akerman recalled with a smile. "He said, 'Michal, do you think you love nature?' And I told him, 'Yes, of course I love nature. I mean, nature is my way of life . . . I'm an agronomist. I learn in the academy. I'm working in agriculture. Of *course* I love nature.'" It seemed like such an obvious question—one that didn't really need to be brought up in the first place. No one asks LeBron James if he enjoys basketball or Thomas Keller if he likes to cook. The answer is simply assumed and embodied in their work itself. "And he said, 'But do you realize that you think you love nature because you deal with agriculture, but if you will think about it, nature and agriculture, they are not coming together anymore.'" He gestured around the vineyard, at the neat rows of vines marching toward the horizon, the perfectly even spaces between them barren of anything but exposed soil. "'Look at this . . . if nature is biodiversity, agriculture is monoculture.'" He paused, then added, "'Listen, you don't really love nature. If you say you love nature, you don't really understand what nature is.'" It was right at that moment, Akerman told me, in a phrase that seemed like it could have come directly from the Old Testament, after some divine revelation or another, "I started to understand."

She looked around at the vineyard that she'd gazed at a thousand times before and realized that the man, whose name she can't recall anymore, was right: this was the opposite of nature. For all the beauty of the neat rows, the clean, pristine appearance of the vines and the way the two arms of each were trained horizontally along the guide wire in what looked like a ballet pose, this wasn't nature as it was meant to be. It was, she began to see, an imposition of human will upon nature. What seemed like a reasonably innocuous question when it was posed, the kind of quip that might constitute small talk between sessions at a soil-sciences conference—*do you think you love nature?*— very quickly forced a recalibration of the entire underpinning of Akerman's understanding of the land and her work. To a surprising extent, it completely shifted the way she saw herself, the way she saw her role in a system that she assumed was working in concert with the natural environment but that had become, over the generations, simply another way of molding the land to fit the needs of the people exploiting it.

She very quickly went down the proverbial rabbit hole, reading everything she could find about what was by then a rapidly spreading movement in the wine world to more seamlessly integrate viticulture into the natural environment. Sustainable farming practices, practical and certified organics, and at the far end of the spectrum, biodynamics, all had been quietly affecting the winegrowing landscape around the world for years. But in Israel at the time, the old ways still tended to hold sway. If Israelis had been able to—as the phrase I heard perhaps more than any other each time I've visited the country—make the desert bloom, then why would farmers (of grapes, of other fruits and vegetables) change the fundamentals of what they'd been doing? It was the very definition of the old cliché that if it ain't broke, don't fix it. (Or the version I prefer, which I seem to recall, correctly or not, the great Jewish philosopher Howard Stern intoning: if it ain't broke, don't break it.) And even if the most dangerous and deleterious techniques and products had been getting phased out for years at that point (Roundup, for example, was on the merciful decline), many of the biggest wine producers in Israel (though, of course, there were exceptions) hadn't yet fully embraced the notions being espoused by SPNI.

I spent a week in Israel during the summer of 2021, and several wine producers pointed out to me, tongue only partly in cheek, that while Europe may be known as the Old World in terms of setting the standard for most of the wines that serve as benchmarks (Bordeaux for Cabernet Sauvignon and Merlot, Burgundy for Pinot Noir and Chardonnay, Champagne for bubbles, and so on), Israel is the Really Old World. More than that, actually: it's the Ancient World. Even today, it's not uncommon to literally stumble across a stone grape-crushing vessel from thousands of years ago in the middle of a vineyard, as I did at Shiloh with Amichai Lourie. With that kind of history, Akerman wondered, why had Israeli wine quality fallen so far behind?

By 2012, Europe had been transitioning to a more sustainable mode of grape-growing and wine production for some time already; in Israel, that kind of work in the vineyard and winery wasn't yet commonplace. It couldn't have been the laws governing kosher production that held the industry back. Despite popular perception, kosher wine laws aren't all that different from the regulations that guide organic production; in the end, *kashrut* is all about cleanliness and respecting the land. But respect for the land hadn't been taken far enough even within those religious strictures. Kosher wine law, for example, requires that a vineyard lie fallow every seventh year to allow it to rest, to replenish itself. But the intensity of the farming that was occurring in the intervening years was so dramatically exhausting the soil that the sabbatical year, or *shmitta*, wasn't enough. The quality of the grapes and soil, and therefore the quality of the wine, was suffering. So, too, was the ability of the vines in much of Israel to live deep into their maturity. Wines made from old vines tend to have more concentration, power, and character than those made from younger ones. In Israel, wine labels with the old-vines moniker were rarer than a bacon-laden brunch in Jerusalem.

Historically in Israel, Akerman told me, "You plant, and then you need to replant. And then you need to replant [again]. And because of that, you are not achieving any great quality." This issue in Israel had nothing to do with the land itself; accounting for the usual and obvious differences in soil composition, underlying geology, and microclimate, Israeli vineyard land always had just as much potential to nurture deep-rooted vines over the course of a decades-long life span. But generations of overfertilizing and overirrigating the land had led to vines that didn't need to sink their roots all that deeply into the earth because everything they needed was near the surface. Not only did this lead to less-characterful fruit but also to vines that were less able to thrive in a dramatically changing environment. So-called old vines, as a result, were few and far between.

As grapevines get older, working year after year to reach the water and nutrients they need, they force their root systems ever deeper into the earth. The best wines, grape growers and winemakers around the world have told me time and again, come from vines that are adequately stressed, struggling just enough to funnel what they need from the subsoil up their roots, through their trunks, and on to the leaves and the fruit. Old vines tend to produce less fruit, grapes which are smaller with more complex skins, and therefore tend to result in more profound wines. But Akerman continued, "[Most] of our vineyards in Israel are, I think, fifteen to seventeen to eighteen years old. That's all. We need to replant because they are getting tired. They are not having . . . high yields [anymore], and we need to replant them." And yet, as she traveled through the great wine-producing countries of Europe, she was being shown around vineyards that were forty, fifty, sixty years old. She began asking herself the obvious question: Why were these vineyards in Spain, in France, and elsewhere dotted with gnarled vines whose tiny berries were downright explosive in character? And why was that not the case in Israel? "It's not making any sense," Akerman told me, exasperated. She added: "It's amazing to see the richness and the full body of the small vine and the small berry. You eat it and you just *chew* it. It's amazing to understand how your mouth is getting full body with a lot of flavors and richness. And here [in Israel], everything is flat. *Everything* is flat [in terms of the flavor of the grapes and the wines before the agricultural transition began]. And I said, 'I want like this, but how can I reach this? How can . . . I grow them for forty years old, for fifty years old?' And then I understand that I can't do it if I'm dealing with artificial. I need to do something with my vineyard. I need to bring nature into my vineyard or to make my vineyard *look* like nature." The very techniques that had allowed Israel's agricultural sector to become such a powerhouse, producing impressive quantities of fruits and vegetables from some of the most unlikely land imaginable, had ultimately placed a cap on the quality of its wines because farming wine grapes isn't like growing other crops: The best producers will almost always be willing to sacrifice quantity to achieve higher quality, and consumers, seeing the rise in expressiveness of a high-end winery's best bottlings, will generally bear the added price burden. That's baked into the system with wine. Good luck doing that at any sort of scale with, say, cucumbers or tomatoes.

So Akerman asked, "How [do] you do it?" How do you grow vines that will last for generations? "You have to add biodiversity. And *natural* biodiversity . . . I mean, you have to do some searching and to understand which are the domestic and endemic trees and bushes. These domestic trees and bushes to bring to your vineyards and to plant them in some spots, and to bring more nature to your vineyard. That's how it happened nine years ago." Little did she know at the time that her work would give her a head start in the country's fight against the ravages of climate change.

Akerman began the process of completely reimagining what a healthy vineyard should look like in Israel—not impeccably neat rows of vines proceeding inexorably toward the horizon with nothing else around to stop them from sucking up as much water and nutrition from the soil as possible but more of a return-to-nature aesthetic, one in which that particular place's natural ecosystem would be allowed to thrive alongside the vines. It would be, essentially, a perfect balance of the domesticated grapevines and a far more natural range of insects, animals, and other plants. This, she reasoned, would not just make for a healthier environment, but it would also facilitate the kind of natural competition for resources that would benefit the vines themselves. And it would, if all went according to plan, lead to vines with longer life spans, similar to what she'd seen in Europe.

When we met on a warm late-July morning in 2021, I was struck by the wild, almost untamed beauty of Tabor's Upper Galilee vineyards; grasses blew in the early-morning breeze, a cacophony of birds chirped and called and sang from seemingly every tree and bush, and the topsoil writhed with beetles and ladybugs and a thousand other critters that I couldn't identify. One of her team members brought up a photo on his phone, a screenshot from one of the night-vision cameras dotting the land that monitors the presence of predators and their prey. It was gruesome: An owl was at the center of the screen, a mouse or rodent of some sort caught in its beak, blood dripping down the feathers below the limp creature. The owl's eyes stared straight at the camera—a blank, almost Hannibal Lecter–like glare. At first, I wasn't sure why Akerman and her team were so proud of this, but when they explained that it meant that a fully functioning ecosystem had been restored to this once-micromanaged land, it all made sense:

the vineyard, finally, was *alive*. It was exactly what she had been aiming for. Today, the wines that are grown on Tabor's land are expressive, incisive, energetic, and often haunting. They also, hopefully, will be better able to withstand the increasingly capricious challenges of climate change, many of which are unprecedented in the modern—and sometimes recorded—history of this ancient place.

On the morning of August 15th, for example—just a few weeks after I returned home to suburban Philadelphia from Israel—wildfires began spreading in the Judean Hills outside of Jerusalem. "It would take three days and the tireless efforts of 204 firefighting crews, 20 planes, IDF rescue teams and help from Palestinian firefighters to fully extinguish the blaze," according to a report in *The Times of Israel*. (The old line *the enemy of my enemy is my friend*, so often applied to Middle East politics and conflicts, apparently applies to climate change too.) In the end, more than 3,000 acres burned—far smaller than the hundreds of thousands of acres increasingly typical in California's wildfires, but a massive swath of land in such a small country, nonetheless.[3] Less than a year earlier, the opposite weather extreme battered the country. Floods washed through several areas of Israel between November 19th and 21st, 2020, and more than nine inches of rain pummeled parts of the country. Based on "statistical calculations, the average return period of such rain intensities is 100 years," the Israel Meteorological Service reported.[4] According to an article in *The Jerusalem Post* from 2021, "Environmental Protection Minister Tamar Zandberg warned . . . that such fires, extreme weather, floods and climate disasters will become more frequent and powerful in the coming years due to the climate crisis. 'This requires us to prepare completely differently for the impending climate disasters,' said Zandberg. 'I am working for the State of Israel to declare a climate emergency. We must define the climate crisis as a strategic threat, which all parties need to prepare for and deal with better. Because from now on it's going to get worse and worse. There is something to be done, and it needs to be done now.'"[5]

Could the work being done by Akerman and her colleagues at the SPNI and other forward-thinking wineries provide a sort of guide map for what that might look like? In other words, could the techniques being perfected by the Israeli wine industry help the country fend off the worst impacts of climate change in general?

Israel is often described as existing in a "tough neighborhood," a reference to the historically fraught nature of the region's politics, the regular spasms of violence, and the constant undertow of distrust between two sides of an argument that has been litigated in countless awful ways for centuries. Despite this popular perception, however, Israel is a country of seemingly impossible achievement, too, and the "Silicon Valley of the Middle East" moniker is well earned and crucially important. Perhaps its most lasting contribution to thriving—or at the very least, surviving—on a warming planet whose weather is growing more and more unpredictable and violent is in the area of agricultural technology. It's important to remember that for all the incredible range of microclimates and geological histories that have shaped it, Israel is still a largely arid land: over 50 percent of the country is classified as desert or near-desert. Remaking that land has been an ongoing project for the state of Israel since its founding in 1948, and the speed with which success was found was unprecedented even in the country's earliest days.

In an article from the April 1960 issue of *Scientific American*, it was noted that "[t]he State of Israel has undertaken to create a new agriculture in an old and damaged land. The 20th century Israelites did not find their promised land 'flowing with milk and honey,' as their forebears did 3,300 years ago. They came to a land of encroaching sand dunes along a once-verdant coast, of malarial swamps and naked limestone hills from which an estimated three feet of topsoil had been scoured, sorted and spread as sterile overwash upon the plains or swept out to sea in Rood waters that time after time turned the beautiful blue of the Mediterranean to a dirty brown as far as the horizon. The land of Israel had shared the fate of land throughout the Middle East. A decline in productivity, in population and in culture had set in with the fading of the Byzantine Empire some 1,300 years ago. The markers of former forest boundaries on treeless slopes and the ruins of dams, aqueducts and terraced irrigation works, of cities, bridges and paved highways—all bore witness that the land had once supported a great civilization with a much larger population in a higher state of well-being."[6]

The authors added that by 1959, just eleven years after it was officially founded: "Israel was already an exporter of agricultural produce and had nearly achieved the goal of agricultural self-sufficiency, with an export/import balance in foodstuffs. It had more than doubled its cultivated land, to a million acres. It had drained 44,000 acres of marshland and extended irrigation to 325,000 acres; it had increased many-fold the supply of underground water from wells and was far along on the work of diverting and utilizing the scant surface waters. On vast stretches of uncultivable land it had established new range-cover to support a growing livestock industry and planted 37 million trees in new forests and shelter belts. All this had been accomplished under a national plan that enlisted the devotion of the citizens and the best understanding and technique provided by modern agricultural science. Israel is not simply restoring the past but seeking full utilization of the land, including realization of potentialities that were unknown to the ancients."

I was raised in a family that like so many of my suburban-Philadelphia neighbors in the 1980s and early 1990s, was more culturally Jewish than anything. Shellfish and pork made regular guest-starring appearances at mealtimes, and our synagogue attendance was mostly limited to the High Holidays in the autumn. Yet my sister and I also attended Hebrew school three times a week (plus required Saturday services) until we each turned thirteen, and Passover *seders* and massive break-the-fast feasts at the close of Yom Kippur were very much a part of the fabric of our home. We had bar and bat mitzvah ceremonies at both our local synagogue and in 1991, atop Masada. And as early as I could remember, my parents gave money to help plant trees in Israel, whether in honor of some important occasion that they wanted to memorialize or because it was a tangible act that we all could understand about where our money was going. Little did I realize at the time that in their small but important way—and I doubt this is a thought that ever really crossed their minds back then—they were helping to revivify the land and, it turns out, inadvertently helping to stave off, just a little bit, the worst impacts of the climate change that would make its most dramatic impacts felt decades later.

The Israeli wine industry has been both a driver and a beneficiary of this; it has, over a remarkably short period of time, become a model for creativity and ingenuity in the face of seemingly insurmountable challenges. And yet for all of that remarkable success, it is now facing climate change–related problems that threaten to undermine everything that has been achieved. Because while reclaiming for agriculture, and more specifically, viticulture, the physical land that once was desert, slaking the thirst of the sand until it supported life, the issues that grape growers and winemakers are now facing have no precedent. And despite the increasingly common threat of fires and floods, it's the less dramatic climate-related impacts that threaten the industry most here.

The 2021 growing season is a good example. By July of that year, Akerman told me, winemakers all over Israel were in touch with one another asking the same question, "What the hell is going on with this vintage?" Because so many of the natural processes that drive a vine through the growing season and coax its grapes to ripeness were just . . . off. Tabor grows grapes all over the country and in a significant range of terroirs and microclimates. Because of that, Akerman had a front-row seat to the climate-change weirdness of the vintage that was affecting the entire country.

In a normal year, grapes go through a series of mile markers toward the end of the growing season, accumulating an optimal amount of sugar as acids diminish, which is called sugar ripeness, and developing a critical mass of flavor and aroma compounds in the skins, stems, and seeds, which is known as phenolic ripeness. In the best years, those two happen at approximately the same time. Picking decisions, then, are based on when the vineyard manager and winemaker feel that a balance between them has been achieved as well as a balance between the acid and sugar inside each berry: Wait too long to pick, and overripe fruit without enough acidity will result in flat, flaccid wines. Pick too early, and the wines will be tart, overly tannic, and lacking in generosity and complexity. Yet in 2021, the received wisdom of the past was no guide for what Akerman (and other industry professionals around the country) were seeing: The vineyards that she manages up north, in the mountains, were experiencing a steep rise in phenolic ripeness by early July, yet their sugars remained perplexingly low for some time after that. Her more southerly vineyards—the ones that lie closer to sea level—were experiencing the opposite: sugars were spiking, yet the phenolic compounds that lead to complexity of flavor and aroma were stubbornly missing. "Everything is weird this vintage," she told me. Still, she went on, "I understand why it's happened. It's all climate. It's because first, this season began after a very hot winter," which didn't allow the vines to have their crucial rest period. The heat meant that they kept on growing leaves and synthesizing the sunlight into sugars when they should have been leaning in the direction of dormancy, which meant that by the time the growing season had gathered speed, their internal cycles were completely off. "And in April," she told me, "it was kind of a freeze wave that made the vines shut down their growing and development and everything changed in the vine." This reversal of the vine's accustomed cycle, brought on precisely because of weather patterns that no one had ever seen or experienced before, resulted in a strange sort of flowering and budding.

In a normal vintage—or every other vintage that Akerman had ever worked—three buds, she told me, form after the flowering: a main one and two ancillary buds that ultimately never become anything. The main cluster is easily identifiable early on as the one that will ultimately develop over the course of the growing season into the bunch of grapes that will be harvested, pressed, and fermented into wine. "But what happened this April, as far as I can see, is that the main cluster just shut down in April in the freeze wave and the small ones just woke up and they're now the main," she said, still perplexed when we spoke in September. Since the smaller buds started off as ancillary ones, they never had the potential to grow into a hearty bunch of fruit in the first place . . . which is exactly what happened. "So that's why we have very low yield [this year] because if you count the number of the clusters per vine, you will see the same numbers as last year. But every cluster, the weight is half [of] what it should be."

A good vintage is one in which the sugar and phenolic ripeness of the fruit occur around the same time and the acid and sugar in each individual berry is in harmony. A great vintage also includes a substantial crop at harvest. Making top-quality wine is one thing, but being a business, volume is needed to keep it viable year after year.

The diminutive size of each bunch of grapes that Akerman observed is a problem that producers all over Israel faced in 2021. "Everyone's speaking about the low yield in Israel this year," she said. "Like 30, 35 percent lower than we predicted" in many vineyards . . . and sometimes more. This problem of diminished volume was a much bigger issue at lower altitudes, where the heat waves of the season were expectedly more severe than at higher elevations, making her losses there closer to 10 to 15 percent. But for many producers who rely on lower-altitude vineyards, the results were financially taxing at best and approaching catastrophic at worst, especially considering that 2022 is a *shmitta* year. For the largest producers, who rely so heavily on a significant volume of production each year, Akerman told me that their only option is to purchase wine or juice from Jewish grape growers elsewhere in the world—for example, Spain, Portugal, and Argentina—and blend it in with their own in a ratio that by law, must stay below 50 percent. And even though this will be done only for the lower-end wines in any given producer's portfolio (e.g., the wines that will be sold for religious purposes or on the proverbial bottom shelf at the supermarket), they risk "losing some identity," Akerman said. The higher end of the industry won't risk that loss of identity because they will remain what are called "wines of place," produced in smaller volumes in 2021 from the grapes they *were* able to harvest, expressive of the unique patch of the planet in which they were planted and grew. But the situation risks tainting the entire industry at just the time that it's finally starting to make major inroads into serious international wine circles.

And it's not just an issue of the 2021 season, although the challenges that the climate caused that year are particularly illustrative. Nothing, Akerman lamented, is predictable anymore, and farmers of any agricultural product rely heavily on what has happened in the past to take full advantage of the future. "It's amazing what's going on," she marveled. "I just spoke with a lot of viticulturists this week about what's going on and . . . it's amazing. No two years [are] identical. I mean, it's something you can't predict. You think you've seen everything, and then . . . surprise! I'm not sure how it will affect the quality, but for sure on the quantity, there's a big effect now. There's a *big* impact." Luckily, the quality of the fruit has been increasing for years now—it's been a major catalyst for the success of Israeli wine in general. But if quantity keeps on going down, that poses real problems for the continued viability of the producers with the least capital behind them. It will force prices to climb even higher than they already have for Israeli wine, and at some point, consumers will start looking to other parts of the world for high-quality yet more affordable bottles. "This is the problem," she said, "because I used to control a lot of things [in the vineyard], and now I feel like I'm losing it . . . that someone is controlling me."

There is, however, a solution, or at least a way to keep at bay many of the most dangerous impacts. "Going to ecological, sustainable agriculture will mitigate," she predicted, "will reduce the climate-change impact, that's for sure." Grapevines that are grown and nurtured with fewer chemical inputs—herbicides, pesticides, and more—have countless advantages over their counterparts. Their deeper root systems more efficiently pull water and nutrients from the underlying geology, which makes them more resilient in the face of both seasonal challenges and more catastrophic ones like hailstorms and fire; they're able to thrive with less water in drought years since they're generally not overirrigated in the first place and have grown accustomed to going without an overabundance of hydration; and so much more. But even a highly naturalistic vineyard program like hers isn't immune to the impacts of climate weirding—just look at her diminished volumes in 2021.

The other option is to begin the process of pivoting away from grape varieties that are no longer suited to the climate as it continues to warm. Growers and producers in up-and-coming wine regions, for the longest time, based what they planted not on what was necessarily best suited to the environment but rather to what they believed consumers wanted the most of. But that's a recipe for mediocre wines at best and terrible ones at worst. "Not every variety [is suited] to every country," Akerman pointed out. "For example, if you ask me ten years from now what will be the main variety in Israel, I think Cabernet Sauvignon will go down, and a lot of varieties like Marselan and Petit Verdot and Grenache and Carignan and Petite Sirah will rise up, and Malbec maybe . . . [but] for sure you're not going to see here Pinot Noir," which needs cooler weather to thrive than Israel is capable of providing, especially in light of the changing climate. The issue there is that outside of professional circles, most consumers just aren't all that familiar with grapes like Carignan and Marselan. How will they be convinced to buy them?

The other side of that coin is how Israeli grape growers and winemakers are becoming more and more open minded when it comes to *where* they're planting vineyard land itself, as opposed to just accepting the old logic that dictated what areas were best suited to the cultivation of wine grapes. Climate change is making this reconsideration even more pressing. Montefiore told me that he thinks of it as "learning to understand the character of a vineyard and to adapt *within* the vineyard, rather than making blends in the winery" to fit a perceived market need.

In a country like Israel, which is finally earning the international wine reputation it deserves, this represents a particularly important development. Because while grape growers and winemakers around the world tend to use past experience and received wisdom as jumping-off points, there is an even deeper well to pull from in this land—for better or for worse. It's ironic, then, that the history of wine here—from the Canaanites and Israelites to the Greeks, Romans, Nabataeans, Byzantines, and now Israelis—has also proven to be one of its Achilles' heels. "Up to now," Montefiore explained, "there's been an attitude [of], 'We've been making wine for thousands of years; we'll make wine for another thousand years.' So it's like a head-in-the-sand attitude in the industry." But that's just not a tenable posture anymore—not in the face of a dramatically changing climate that manifests its shifts in a hundred ways both predictable and, increasingly, catastrophic.

Yet Israel is perhaps better positioned to survive these changes than many other wine-producing countries, for reasons both historical and cultural. Pivoting in the face of dire threats is what the state itself has done since its founding and what the Jewish people have done for thousands of years. This is why I shouldn't have been surprised when during my most recent visit, so many producers spoke of an increasing focus on growing grapes in the desert, which I learned, while completely counterintuitive at first, is actually a wholly logical throwback to the viticulture of the ancients.

The desert, after all, may seem like a terrible place to plant vineyards. But in Israel, at least, it actually has a lot going for it. The days are hot, of course—it *is* the desert, after all—but the nights are predictably cold, which provides an excellent diurnal swing. And within Israel's borders, the desert regions provide a good opportunity to gain altitude. Montefiore pointed out to me that while more than 50 percent of Israel's land is technically desert, it's those areas that tend to be relatively mountainous, the verdant Golan Heights notwithstanding. Modern winemaking wisdom and old-fashioned common sense, of course, implied that growing *anything* in the desert is a questionable idea and sure to be fraught with challenges, but Montefiore told me, echoing a sentiment that I've heard countless times during my visits, "Israelis are stubborn, crazy people. And [if] you say to them, 'You can't grow wines in the desert' . . . there'll be some people that will do it just to show you [they] can. And we now have something like 5 percent of Israel's vineyards growing in the desert. The desert is over 50 percent of Israel in size, depending where [you consider] the borders of Israel, so you can never say exactly. So I'll say over 50 percent, which is an enormous area. And to fly down to Eilat [in the far south of the country] . . . and you look out the window and you see a square bit of green and it's a vineyard. It's actually quite moving. Or if you go to Mitzpe Ramon," a town perched at the precipice of the Ramon Crater in the Negev Desert, approximately 2,400 feet above sea level, "And you see like a green river of vineyards running through an old, [arid] river valley. It's amazing."

The move to more elevated desert sites has had a number of unexpected benefits. "It's high altitude; it's very cold at night. It can be misty in the morning, which protects the grapes. There is no humidity [otherwise], there is no vegetation, so the threat of disease is very low," Montefiore went on, sounding genuinely excited. "And this is proving to be a genuinely good area, particularly for white wines [in] the Negev." White wine in the desert! It's a concept that flies in the face of so much received winegrowing wisdom that it almost sounds like a parody, a cautionary tale—something an old-school Jewish comedian would have joked about at a 1950s Catskills resort. But the results in the glass are beyond compelling.

Growing grapes in the desert does, however, come with some unique challenges, according to Montefiore: "Recently, in an article where I was writing about the Negev, I said how throughout [most of] Israel, the big pest is the wild boar. The wild boar comes to eat the vines when they're ripe. They know when they're ripe," he explained. It's an issue in some of the greatest wine regions of Europe too; in Tuscany, for example, wild boars, called *cinghiale* in Italian, are a scourge. But Montefiore went on, "In the Negev, it's camels that come and eat the grapes. And they'll eat a vine to the floor like a salad, and eat it all. And then no one says they own the camel; it's owned by [a] Bedouin. And of course, it's not like a dog with a chip, so you can't find who owns it. So they can cause a lot of damage, camels." Looking on the bright side, he paused and added, "This is certainly a unique selling point! I don't know how many regions around the world have camels in their story. So that's the Negev."

This move to the desert, and to the higher altitudes it often provides, is also giving winemakers and grape growers the ability to experiment with a greater range of grape varieties than they might have been able to plant in more traditional locations. While climate change is forcing a reconsideration of where to plant wine grapes, the conditions are, in some cases, simply better in the desert mountains for successful vineyards than they are in the more traditional winegrowing areas closer to the coast and lower in altitude. There are plenty of great wines being made in a wide range of environments and altitudes throughout Israel, but it's impossible to ignore the benefits of reconsidering the received wisdom. "There's been an enormous change in the quality of Israeli wines, and the quality of agriculture [for] Israeli wines, in the last thirty years," Montefiore explained. "And there's been enormous change in the importance of the vineyard. The Israeli vineyards were [historically] planted in the coastal regions. Baron Edmond de Rothschild [who was an owner of Bordeaux's Château Lafite and founded the modern Israeli wine industry in 1890 when he built Carmel Winery, which still uses the original structure for some of its production] planted [vineyards] where the farms were," because that's what seemed to make the most sense back then. "So you had your village, you had your cows, you had your crops. You planted your wheat; that didn't work. You planted your potatoes because you needed [food], and *it* didn't work. They planted grapes, and it works." This planting of grapes where other crops often failed, in fact, is an ancient way of doing things. Aside from the fact that apples and grapes tend to grow well in the same or similar spots, good wine grapes almost never thrive in the same conditions, and in the same soils, as many other crops.

"So these [early Israeli vineyards] were in the coastal regions of Zikhron Ya'akov and southeastern Tel Aviv," Montefiore continued. "And it was only later [that] the Golan Heights Winery was the first winery to plant at altitude and to understand that you had to manage the vineyards to control the quality. That was the first time that the decisions of what to do in the vineyards reverted from the grower to the winemaker." That change, from the grower to the winemaker making many of the calls in the vineyard, often marks an important shift in wine quality all over the world when it happens; it means that the process of sacrificing quantity to achieve greater quality has likely begun, which is often a prerequisite for world-class wine. (As long as the winemaker isn't dictating changes to support stylistic demands that go against the character of the land.) This first happened at a notable scale, according to Montefiore, at Golan Heights Winery. "And since then, the wine industry moves northwards and eastwards in search of higher altitude," he said. So while it's counterintuitive at first glance for the wine industry of a country that's more than half desert to shift its focus from the sea, from areas that ostensibly have greater and easier access to water and more moderate conditions, to regions that are landlocked, more naturally arid, and seemingly hotter, the results have been impressive: Israeli wine, especially in these challenging locations, is regaining the quality that it's supposed to have had thousands of years ago. I've personally been impressed.

The Israeli wine industry has also undergone a serious process of education, and its learning curve has been remarkably short largely because of an increasingly worldly point of view among those who make and grow the wine. Golan Heights Winery was the first in the country to hire a winemaker from the University of California, Davis, the Harvard of wine-education programs, in the 1980s. (Unfortunately, he stayed for only a year before homesickness, and the remote location of the winery, got the better of him, and he left. But his impact was huge, as was the fact of his willingness to work in Israel at all given the reputation of the country's wines at the time.) As is the case in the wider world of winemaking, Israeli winemakers started to not only *look* outward but to travel the world, harvest grapes, and help make wine in regions from Napa to Bordeaux to Australia and beyond. "What we have now in Israel is an incredible wealth of knowledge," Montefiore explained, "because people are trained everywhere. Anywhere there's a university for wine, they study. They come back to Israel, having done studies in those countries, [and bring those ideas, those techniques]. And so we have a very youthful, knowledgeable winemaking nucleus who bring with them a sort of melting pot of ideas from around the world." As recently as 2020, Ido Lewinsohn, of Barkan Winery, became just the second Israeli Master of Wine (the first was Eran Pick of Tzora Vineyards). And his work at Barkan has been incredible; while the brand built its reputation on widely available kosher wines that are familiar to Jews around the world, he has placed an increasingly passionate focus on also crafting smaller quantities of unique, idiosyncratic reds and whites that have garnered high scores from critics and serious admiration among sommeliers. It's telling that someone who holds the prestigious title of Master of Wine—there are 418 in the world—would choose to work at such a large producer like Barkan. But the capital and distribution power of the brand itself, its facilities, and its American importer allow him to push the envelope and to do things he might not have been able to with a smaller producer behind him. (He makes his own wine under the Lewinsohn Garage de Papa label too.) Producers all over the country are taking important steps toward higher quality, toward more sustainable farming and production. Tzora Vineyards has been certified by FAIR'N GREEN. Golan Heights Winery is certified by the LODI RULES Sustainable Winegrowing Program. "Galil Mountain is another winery pioneering sustainability," Montefiore noted. "Tabor, Golan, Galil, and Tzora are leading the way."

It may have taken thousands of years, but the Israeli wine industry is finally recovering some of the glow that it had back in the time of the ancients. Sommeliers and critics around the world are increasingly giving it the respect it deserves—and in fairness, the wines are earning their new status in ways that they hadn't for far too long. Vintage after vintage, the quality of Israeli wine seems to be improving, even if climate change is occasionally posing serious challenges to the quantity, as in 2021. In just the nine short years between my tasting-focused visits of 2012 and 2021, the character of the wines completely shifted. Back then, the supposed best of them possessed plenty of power and density but far too often fell into the trap that premium bottlings did around the world at the time: they were often too ripe and powerful, with an overlay of oak that wasn't just unnecessary but that in many cases, covered up the character of the land itself. There were exceptions, of course, but they were few enough that they more or less proved the rule. Fast-forward to 2021, and the situation had reversed itself: the best wines were built on the same philosophical foundation that most of the greats of the international wine world generally are—that wine is a lens through which to see a particular piece of the earth, and the jobs of the grape grower and winemaker are to fine-tune the focus of that to bring it into ever sharper clarity. Respecting nature and integrating winegrowing as seamlessly as possible into its systems not only results in better wine but in a healthier overall environment and a greater ability to weather its storms, proverbial and actual.

Israel's work to not just combat the impacts of climate change on its wine industry but also to find a way to thrive despite them has gotten the attention of some of the most important people in the business. Last year, on a trip that was so secret that everyone involved had to promise not to divulge names, the principals of one of France's most prestigious producers spent a week in Israel speaking with growers, soil scientists, academics, winemakers, and more in an attempt to understand how Israel was dealing with climate change. The symbolism wasn't lost on anyone. Israel's modern wine culture was kick-started in the late nineteenth century by Baron Edmond de Rothschild, who brought Old World know-how to this ancient place. Now, 130 years later, the French wine elite were back, this time to *learn* from the Israelis. It was the old line about the student becoming the teacher brought to life.

Unfortunately, the changing climate is constantly throwing more and unexpected challenges at Israel, threatening all of this hard-earned success at just the time that its wines are gaining international traction. Passionate and determined professionals around the country are experimenting with and implementing practices in the vineyard and the winery to allow them to stave off the worst of the effects, but catastrophic climate events like fires, freak freezes, floods, and more are occurring with greater violence and more frequency. Still, Israel isn't known as a "start-up nation" for nothing, and its technological and agricultural advancements, and the willingness of its people to find solutions to even the most menacing problems, are woven into its culture. The country's wine industry has as good a chance of surviving, at least in the short term in our now dramatically changing climate, as almost any other I've visited. What the long-term prognosis will be, however, is anyone's guess. Israeli technology and tenacity have helped raise the industry up to impressive heights in a remarkably short period of time. But there is a limit to how far it can go in the face of a warming world, which, no matter what, always seems to win in the end. "Who knows if in a hundred years' time Israel will be a country for making wine," Montefiore told me. What that means for the rest of the wine world, too, is terrifying.

4

A BRIGHT FUTURE FOR OVERCAST ENGLAND

The Judgment of Paris wine tasting in 1976 pitted the best of Bordeaux and Burgundy against a selection of Cabernet Sauvignon and Chardonnay from an upstart American wine region called . . . wait for it: Napa Valley. If you've read the excellent book by George Taber, *Judgment of Paris: California vs. France and the Historic 1976 Paris Tasting That Revolutionized Wine*, or seen the less accurate but still highly entertaining movie *Bottle Shock* (which was loosely based on the book), then you know that what started off as a lark and provoked a deep sense of ambivalence among the French winemakers and other wine professionals who judged the competition (*how*, they wondered, *would any wine region in the world stand up to the best of France, much less one from—gasp!—the United States?*) ended up shaking the very foundations of the wine world. Because the Americans, after the bottles had been emptied and the spit buckets filled, ending up winning among both the reds and the whites. The Stag's Leap Wine Cellars S.L.V. Cabernet Sauvignon beat out legends like Château Mouton Rothschild, Château Haut-Brion, Château Léoville-Las Cases, and Château Montrose; and Chateau Montelena topped the list of Chardonnays over icons like the Joseph Drouhin Beaune Clos des Mouches, Domaine Leflaive Puligny-Montrachet

Les Pucelles, Ramonet-Prudhon Bâtard-Montrachet, and Roulot Meursault Charmes.

The French had been producing what most professionals considered to be some of the best wines in the world for so long that it never occurred to anyone that their supremacy would (or could!) ever be challenged, certainly not by some region in California. But that's exactly what happened, and now, nearly half a century later, the California wine industry isn't just one of the most prestigious in the world but also one of the most valuable. In August 2021, *Barron's* reported that "within the last 10 years, California's share of the global wine trade has risen to 7.1% from 0.1%, making it the most important region of the world outside of France and Italy."[1] The report added that among California's so-called cult wines (prestige reds made in small quantities largely but not exclusively in Napa Valley; among the most famous are Screaming Eagle, Harlan Estate, Dalla Valle Maya, and Sine Qua Non, among others), values jumped 29 percent in the two years between the middle of 2019 and 2021. By comparison, First Growth Bordeaux went up 17.5 percent and red Burgundy, just slightly more, at 19 percent—still significant but far outpaced by California. An industry that started out as an underdog has become a standard bearer not just in the United States but around the world.

I bring this up because there are fascinating parallels between the world of California reds and whites in the 1970s and English bubbly right now. In fact, "English fizz," as it's affectionately called, even had its own victory at what may, in the future, go down as that country's equivalent of the Judgment of Paris.

In 2015, *Noble Rot* magazine (the name, though it sounds like a punk band from Manchester, is actually a reference to the fungus that affects certain grapes, dehydrating them and changing their flavors, resulting in some of the most ambrosial wines in the world, among them the great sweet whites of Sauternes and the impossible-to-pronounce-unless-you've-had-a-whole-bottle Trockenbeerenauslese from Germany) organized a blind tasting that pitted sparklers from England against top names in Champagne. The judges included Jancis Robinson, MW, one of the most respected wine critics and writers in the world—she was even awarded an Order of the British Empire for her work—Jamie Goode, a wine author and highly influential wine writer for many top publications; and Neal Martin, who back then was a critic for Robert Parker's *Wine Advocate* and is now senior editor and a critic at *Vinous*. Two English sparklers, the Hambledon Classic Cuvée and the Nyetimber Classic Cuvee 2010, bested Champagnes from Pol Roger, Taittinger, Savart, and Veuve Clicquot, among others. Everyone was shocked, including, perhaps inevitably, the English.

At that time, English sparkling wine had been on the upswing for years. Today, its proverbial stock has risen even higher, gaining not just market share but also the attention of more and more retailers, restaurant wine buyers, and sommeliers, which if the trend continues, will likely lead to an increasing willingness to purchase among consumers themselves. It's already starting to happen. And while English wine has historically been looked at with approximately the same level of enthusiasm and respect as the country's culinary history, there have been improvements bubbling (sorry!) beneath the surface for a decade or more now. (The English culinary scene is vibrant, exciting, and delicious too.) Much of this is following a pattern that would be familiar to professionals in once-overlooked (or even disrespected) regions around the world. Grape growers are focusing more on quality than quantity, planting the varieties that are best suited to a particular soil type and microclimate rather than what they perceive the market as wanting, and winemakers are vinifying that fruit in ways that are far more driven by qualitative than financial considerations than ever before. They're also bringing the experience of working in vineyards and wineries around the world back home. As quality goes up, so does the reputation of the wines, which allows them to be sold for more money. This, in turn, reinforces the benefits of improving their quality and making the investments necessary to continue the direction of those trend lines. It becomes a positive feedback loop, a virtuous cycle of sorts, and regions from Mendoza, Argentina, and the Central Valley of Chile to the vineyards perched on the flanks of Mt. Etna and, yes, planted within the oil painting–worthy hills and fields of Southern and Eastern England have benefited.

And as far as England goes specifically, there is a legitimate argument to be made that much of this success is thanks, in no small part, to climate change.

Contrary to popular belief, there is nothing inherently inferior about the terroir of southern England, where much of the country's sparkling wine is produced. In fact, the opposite is true. In the southeast, home to the important winemaking counties of West and East Sussex, Kent, Surrey, and Hampshire, the often chalky soils there—joined or interspersed with clay, sand, and loam—bring to mind the many terroirs of Champagne. Both southern England and Champagne are relatively cool-climate wine regions too. (Tourists generally don't go to either place for a tan.) And Sussex and Kent, which are home to a number of top-quality sparkling wine producers, are a lot closer to Champagne than most people realize. Sussex, for example, is only around 250 or 300 miles from Reims, France, the epicenter, along with Epernay, of French bubbly. Interestingly, both southern England and Champagne are part of the geological formation known as the Anglo-Paris Basin, which has its origins hundreds of millions of years ago. At the risk of oversimplifying, the geology of Champagne shares similar origins with the geology of southern England . . . it's just that the English Channel and Northern France get in the way. And today, as temperatures rise at the same time as grape-growing acumen and winemaking skill do, it only makes sense that English fizz is finally having its moment. Then again, a burgeoning English wine industry spurred on by a warming planet is nothing new. According to an article in the *Los Angeles Times*, "England is, in fact, not a new wine region at all. The Romans practiced viticulture there, and there's evidence of a heyday about 800 years ago, when the last great epoch of global warming overtook the planet. But the current warming trends have gotten vignerons to start planting again."[2] And they're doing so with gusto.

Of course, there are several key differences between England and Champagne, not least of which are the respective climates in each place. Champagne has historically been slightly warmer than southern England, and this has allowed grape growers to ripen their fruit with more regularity. That's not to say that ripeness has been reliable there; far from it. The main reason that the flagship bottling of the vast majority of Champagne houses is the brut non-vintage is that it has served as a hedge in those years when it just wasn't warm or sunny enough to harvest grapes that were adequately ripe to produce high-quality vintage wine. And since non-vintage Champagnes are crafted from both a current vintage and a wide range of what are known as reserve wines, Champagne houses have been able to release a consistently styled (and consistently available) wine year after year regardless of the character of the vintage on which each new one is based. According to Vinepair.com, "Not only do winemakers in Champagne need to create a product each year to survive, they also need to create a consistent product, thanks to Champagne's renowned reputation in the global market. To combat this issue, Champagne houses create most of their cuvées as NV, or non-vintage. This signifies that Champagne producers use grapes and must from various vintages to create their cuvées, ensuring its consistency. In other words, this means that each specific producer has a distinct formula comprised of various vintages, resulting in steady flavor profiles year after year."[3]

That's the genius of the *chef de cave*, or cellar master—the ability to blend together a different range of lots each time that will result in a bottling that tastes more or less the same as the one before and the one before that. These days, however, climate change is shifting how the sparkling wines of Champagne are being farmed and blended. While warming temperatures mean that vintage Champagnes are able to be produced more frequently, and less sugar is generally being added to them all before bottling (a practice called *dosage*), climate change is also causing serious challenges. If trends continue as they are, temperatures could become a bit *too* warm for Pinot Meunier, one of the three main grapes in the classic Champagne blend (Chardonnay and Pinot Noir are the big two). The little-known Petit Meslier grape variety could possibly replace it one day in Champagne blends; it possesses enough acidity to remain fresh even in the warmest years.

And in some years, the problems are even more dramatic: In 2021, according to a piece in *Food & Wine*, "The Comité Champagne—the appellation's official trade association—announced that as much as 60 percent of the region's yield may already be lost due to this year's poor weather conditions. Specifically, the Comité Champagne says that the 12 days of early frost in April likely cut yields by 30 percent, followed by 'persistent rain' throughout the spring which allowed mildew to claim another 25 to 30 percent. Making matters worse, the region was hit with hail on 'several occasions,' damaging over 1,200 acres of vineyards, with half losing their entire crop — though thankfully that represents less than one percent of Champagne's total vines." Still, the piece continued, "despite all the damage, optimism prevailed as the harvest got underway, sticking with the familiar mantra that a drop in quantity will not mean a drop in quality."[4]

Yet while a warming climate is challenging Champagne in many ways (though quality these days is magnificently high), it is, on the whole, benefiting southern England. Even farther north, in East Anglia, wine is a growing business too. Though the first vineyards were only just planted in the second half of the twentieth century there, the region is generating a substantial amount of buzz. And while southwestern England, which has more clay, along with chalkier soils in the eastern part of the region, and more rocky soils with a notable hit of slate and granite as you head farther west, may not closely resemble Champagne geologically, its wines are excellent. These regions, which with southeastern England together account for the vast majority of England's vineyard lands, are benefiting tremendously from warming average temperatures, which aren't just allowing for greater ripeness but also for a better selection of grape varieties than could have been planted in the past.

"I suppose we *have* got warmer because otherwise we wouldn't be growing . . . stuff like Chardonnay and Pinot Noir," noted Bella Spurrier, who founded Bride Valley, the highly regarded sparkling-wine producer in Dorset, with her late husband, the wine icon Steven Spurrier. Ironically, it was Spurrier, founder of the legendary Paris wine shop and school Académie du Vin, and his business partner, Patricia Gastaud-Gallagher, who conceived of and put on the 1976 tasting that would become known as the Judgment of Paris. "It used to be," Spurrier continued, "that the English growers only grew hybrid grapes," which are varieties that have been created with a specific set of characteristics in mind, among which the ability to give great pleasure isn't necessarily at the top of that list. Surviving and thriving in a particular place, however, is. When the Spurriers founded Bride Valley in 2009, they were confident enough in the Dorset climate to plant the classic Champagne varieties—Chardonnay, Pinot Noir, and Pinot Meunier—just as other top producers of English fizz were doing. Hybrids would not be part of their plan.

The results have been impressive, and the wines have garnered a great deal of praise in the international press. That doesn't mean that it's all been easy, of course, because as farmers, distillers, and winemakers around the world have told me time and again, climate change is not just about temperatures getting warmer on average but also about the utterly unpredictable effects of those climbing temperatures—the nature of precipitation, drought, and more. But England's location and the effects of the sea have proven to be beneficial in this regard too. Climate change, Spurrier told me, has impacted both England and France but in very different ways. France, she said, "has it more [extreme] because [parts of] France [have] a [largely] continental climate." Champagne specifically, according to the Comité Champagne, has a "dual climate, which is predominantly oceanic but with continental tendencies."[5] "We have a totally maritime climate," Spurrier continued. "Nowhere in England is more than a hundred miles from the sea, so we're affected by the sea wherever we live in England, I think, and certainly we are here. Here at Bride Valley, we're only two miles from the sea or the coastline, so we are very much affected [by it], and we've benefited in many ways because it prevents us from getting such bad frosts as many people get inland." Still, it's not all getting easier. "Certainly, we've had more vicious, harder rainstorms and gales," Spurrier lamented. "I mean, I grew up in England, and I don't remember ever having such gales and rainstorms as we have now."

In general, however, despite the challenges, English sparkling wine has benefited from climate change—it just hasn't all been as cut and dried, so to speak, as popular myth has it. Still, as the potential for growing high-quality grapes continues to increase, more and more vineyard managers and winemakers are gaining greater knowledge about what they have to do to produce top-quality wines. As that has happened, the market for them has grown. "I think that the quality has been going up and people have stopped thinking of English wine as a joke because it really is okay." Spurrier added with her characteristic understatement, "Nothing like a bit of confidence to get people going, isn't there?"

In fact, conditions at Bride Valley have gotten to the point that the team there has been able to produce still wines in warm years—not just the relatively easy-to-work-with Chardonnay but also the famously finicky Pinot Noir, which Spurrier said, "is perfectly acceptable. It isn't a Burgundy, but it's perfectly acceptable Pinot Noir, and nobody would've thought of that a few years back." Continually improving Pinot Noir, grown and vinified in England? That's high praise, indeed.

The climate-change story that's generally been told about the English sparkling-wine industry in particular—and the entire English wine world in general—goes something like this: climate change is causing average temperatures to rise, which is allowing grape growers and winemakers in England to plant better grape varieties, harvest them at optimal ripeness, and finally produce wines that can rival those of Champagne, albeit in their own style.

That's not entirely true, however; climate change is such a complex phenomenon that attributing any one wine region's—or country's—growing success to its most well-known effect is bound to be inaccurate at best and misrepresentative at worst. Adam Williams, sales director of Balfour Winery, a highly regarded producer in Kent, certainly thinks so. "Obviously, English wine is changing a lot, and the climate is a big part of that," he told me. "Often, [however], you read that that's the only reason that English wine is [improving]." The truth, he said, is that "it's usually oversimplistic to say that, because there's lots of other reasons." Among them, he pointed out, are improving clonal selection, site selection, and investment—the same factors that impact wine regions around the world. The difference in England now is that climate change is making certain decisions possible that simply weren't before.

Clonal selection is exactly what it sounds like: the choosing of which particular clone of a grape variety to work with. Even within one particular grape variety, there are multiple clones that boast different flavors and aromas, have affinities or problems with too much water or the lack of it, tendencies to grow smaller or bushier canopies of leaves, and more. Planning a vineyard is about more than simply deciding which grape varieties to grow; it's also about figuring out which versions, or clones, of those varieties will do best in that particular climate and in those particular soils. There's also the issue of rootstock, since most wine grapevines aren't planted on their own roots but rather grafted onto specifically chosen stocks that are best suited to that particular patch of the planet and are resistant to the various challenges that they'll face there, whether fungal diseases, viruses, droughts, pests, or a million other potential issues. Site selection is the art and science of figuring out exactly where to plant a vineyard. Classic wine-growing regions benefit from centuries or more of experience—real-world trial and error, in effect, or the accumulated knowledge of the generations. Younger wine regions, however, don't have that advantage, and England is a great example. Finally, the investment that Williams mentioned is self-explanatory: growing high-quality grapes and making fine wine are expensive endeavors. The more money that's invested in them, the better the end results are likely to be. Money is never a guarantee, but it's an awfully good hedge. Still, all of that having been said, he pointed out, none of those key factors would result in the exploding quality and reputation of English sparkling wine these days "without the fact that we are warmer than we were thirty years ago."

Warming temperatures cannot be ignored, and their impact, even given the massive variation from one year to another (more on that soon), is clear. EnglishWines.info, which is arguably the most comprehensive online repository for the nitty-gritty of the world of English wines (it's run by Tony Eva and was recommended to me by Adam Williams) ran an analysis in 2015 called "Growing Season Heat Accumulation in Central England Since 1950," which shows this trend line quite clearly. (Though the heart of English sparkling-wine production is in the south of the country, there are more and more producers taking advantage of both the terroir and temperatures northeast of London, in East Anglia, too.) Using the wine-industry standard metric of growing degree days (GDD), which quantifies the accumulation of heat during the stretch of time that grapevines mature from budbreak to harvest, and admitting that there is a tremendous amount of vintage variation within the forty-five-year span of his analysis (this is England, after all), Eva writes that, "within the annual variation there is a significant underlying trend which shows an increase of 175 GDD units [347 F] since 1950. As a result of this upward trend, GDD in the past 20 years for central England frequently exceeds 800 GDD units [1472 F], whereas previously this only occurred in exceptional hot years such as 1959, 1976 and 1989."[6] Numbers like that mean nothing in a vacuum, so we need a sense of context. In 1950, Central England had around 1200 F GDD; by 2015, it averaged more than 1400 F. Champagne is generally just over 1800 F, which goes a long way toward explaining why English growers are increasingly coaxing high-quality fruit to maturity before it's transmogrified into bracingly fresh sparkling wine.

Again, however, even the best fruit wouldn't make much of a difference—or for that matter, *any* difference—without grape growers and winemakers who know what to do with it. And in that regard, things just keep on improving: English sparkling-wine know-how is very accomplished and constantly on the upswing. So, too, does the market for these wines, and much of that has to do with both the increasing quality of the wine and the growing focus among consumers when it comes to eating and drinking locally and supporting their local producers. Williams told me that he was recently on a call with a market-tracking agency that looks at the broad trends shaping how and what consumers eat and drink around the world, and the consistency across nations was remarkable. "[The Spanish] are drinking more Cava, Italians are drinking more Franciacorta, and the English are drinking more English sparkling wine," he told me. "So this drinking-local thing is a global trend within sparkling wine," and the ramifications are huge. As a local consumer base begins supporting the producers in their proverbial backyard, those same producers make more money, which is often plowed (pun most definitely intended) back into the growth of their grapes and the production of their wines. It's dominoes falling in the best possible sense.

"I'm fascinated by the amount of records that we're having at the moment in terms of recorded temperatures in the UK," marveled Dermot Sugrue, one of the most important winemakers and wine consultants in England right now. "[The year 2021], I think it has exceeded last year . . . as the hottest temperature ever recorded in the UK. I know that in the last five or six years, '15, '16, '17, '18, the temperature in the UK has been, on average, warmer every single year."

Sugrue has been making sparkling wine in England since 2003—ironically, a year whose heat waves made news around the world, led to thousands of deaths across Europe, and resulted in wines that especially in Bordeaux, proved to be both very controversial and wildly popular commercially thanks to their opulence, ripeness, and generally higher alcohol. Before then, Sugrue had graduated university with a degree in environmental science and somehow found himself working as an advisor in Richard Branson's financial company. After a time, he realized that the financial world just wasn't for him, so he made a transition that countless others had before. "When I was about twenty-eight years of age, I thought, *What am I doing? I really hate this*. And I loved making wine. I started making beer and wine when I was a teenager in Ireland, so I decided to go to Bordeaux. And I did a couple of vintages in Bordeaux and then came back to England," he told me. "And I realized what was going on in 2003, in England, because suddenly, the sparkling wines being made in Sussex since the 1990s had matured and were all released. Journalists and critics all over the world were talking about these wines, and particularly the Nyetimber wines." The Nyetimber estate, which was referenced in 1086 in the Domesday Book, saw its first modern vineyard planted in 1988. Even then, the potential for Chardonnay, Pinot Noir, and Pinot Meunier was clear . . . though at that time, successfully crafting world-class sparkling wines in England was a far more tenuous endeavor than it is today, and it took the right people to see it. As Neal Martin wrote in *Vinous*, "As any chef will tell you, you need quality ingredients to make a tasty dish, and our ingredients just did not cut it. Enter two Americans, Stuart and Sandy Moss, who saw the parallels between our white chalky soils and Champagne's white chalky soils and thought: 'Hang on a moment. . . . Those Champagnes taste OK. Why don't we try using the same grape varieties?' So, in 1988, they planted Chardonnay, Pinot Noir and Pinot Meunier in the lee of the South Downs. Inspired by the locale recorded as Nitimbreha in the Domesday Book, the Mosses named their estate Nyetimber. Guess what happened? The wine tasted delicious. It began winning awards and, would you Adam-and-Eve it, beating Champagne in blind tastings. And they have never looked back, give or take the occasional washed-out season."[7]

As Sugrue told me, Nyetimber's sparkling wines had really come into their own by the 1990s, and he was ready to go all in. "I made sure that I got a job at Nyetimber," he recalled. "I got a job as assistant winemaker, and within one year I was promoted to the head winemaker's job at Nyetimber. . . . It was effectively the best winemaking job in the country."

In 2006, he left Nyetimber "in order to start the winemaking project at Wiston Estate," he explained. "And in the same year [I] planted another vineyard for my own project, Sugrue South Downs. Wiston has gone on to be awarded UK Winery of the Year in 2018, 2020 and 2021, while Sugrue South Downs has been awarded UK Boutique Producer of the Year twice, in 2020 and 2021." Early on, he noticed that "there were a lot of vineyards being planted in the U.K. [to] Chardonnay, Pinot Noir, Pinot Meunier, all kind of following the Nyetimber and Ridgeview model, but there weren't cellars appearing," he explained. "There weren't wineries." He decided that the time was right to leverage the good reputation he'd built up and begin offering "contract winemaking services specifically to vineyards who didn't have their own winery, but were looking to make high-quality sparkling wine from those varieties." Sugrue was the first contract winemaker in the country focusing exclusively on traditional-method bubbly, and the experience he's gained has given him a unique perspective of how climate change has impacted the English wine industry as a whole.

"One of the main benefits for me has been the experience I've gained over that kind of decade and a half. [It] has been extraordinary because I'm not making two, three, four wines a year. I'm making 20 to 25 or sometimes 30 individual cuvées in a year. So that's a lot of wine," he pointed out. "I haven't actually counted how many in total, but it's probably in the region of over 200, 250 wines, something like that." Which is all to say that Sugrue has not just seen the shifts brought about by climate change—he's viscerally experienced them. "I think the most important thing really, or the most clear trends throughout, say, the '70s, throughout the '80s, throughout the '90s, the noughties, have been the number of days during the growing season where the temperature . . . exceeds 30 degrees centigrade," he said, or 86 degrees Fahrenheit. "If you look at a simple chart of the kinds of those [temperatures], you know the 1970s there's been a few, in the '80s there's a few more. In the '90s, there's quite a few more, and it is absolutely a consistent trend. Now, as you know, grapevines are a very good barometer of small changes in temperature and changes in climate over time, and I've certainly seen that in the 15 years or so that I've been making."

In recent years, the benefits of climate change have been plentiful. "We've had a series of vintages just in the last four years" that have been quite good, Sugrue said, "with 2018 being a massive year in terms of [most] vineyards in the UK producing three times their average yield. It was extraordinary." The grapes in 2018 were also exceptionally ripe, he told me, which, combined with their abundance on the vine that year, are "two things you don't expect: massive production and perfect ripeness. How does that happen? 2018 was that kind of perfect year, and the fruit was harvested relatively early. 2019 we had another huge year; some very good weather at flowering to give a very good fruit set—another bumper year. But it was a little bit compromised by rain at harvest time, but it was still a relatively early vintage," which indicates that the grapes achieved their optimal ripeness in a relatively short period of time, which is generally thanks to increased heat. "Then 2020, last year, was probably the best winemaking fruit that I've had since 2003 or since 2011." It was hit after hit after hit. Of course, desirable conditions can't be relied on anywhere, much less in England. The year 2021 was a return to more challenging weather. When we spoke in November, Sugrue told me that "we're having a return to the old days when everything was late, and we had to wait and wait and wait and wait right till the very end of October, and actually into the beginning of November, to get the ripeness that we're searching for" before harvest was possible.

Climate change, of course, is also about abnormally cold or wet weather too, and England hasn't been immune to that either. "In September 2011, we had [one of] the warmest recorded temperatures in meteorological history in the UK in over 140 years," he recalled. And then, the following year, "2012 was officially the coldest, the wettest, and the darkest year in England, the darkest summer in England for 100 years, since 1912. . . . So you have the extraordinary contrast, and I guess this must be part of what we're now calling 'global weirding.'"

For all of that weirding, however, Sugrue is optimistic about the future of sparkling wine in England. "I do actually consider myself very lucky to be making wine in England at this time," he told me. "I seem to be the right guy in the right place with the right attitude. . . . It is remarkable what we can make in England. It is absolutely remarkable. But," he cautioned, "I think we need to recognize our limits as well, you know? In 2020, last year, which was fantastic ripeness, for the first time I made a still Chardonnay—actually Chardonnay and [the hybrid grape] Bacchus. And for the first time, I made a still Pinot Noir, a red Pinot Noir, 100 percent in barrel. And these are wines that I'm really, really, really proud of. I don't want to try and make, or plan even, to make another still wine until we know that we've got very, very high levels of ripeness. And for me, then, the perfect thing is to make sparkling wine every year, and then only when we have got exceptional conditions, *then* try to make a still wine, [or at least] *try* to make a still wine." Which is a statement that would have been considered absolutely crazy twenty-five years ago, much less from the country's preeminent sparkling winemaker.

There are still plenty of challenges ahead—growing high-quality grapes and producing world-class wine is a perplexing endeavor anywhere, to say the least. Even in the most highly regarded regions, in the most promising vintages, it's impossible for anyone involved in the process to let their proverbial guard down until the wines are safely in their bottles, sealed beneath their corks. And even then, in some parts of the world, risks remain. Wildfires in California, earthquakes in Chile, and more have claimed countless bottles and barrels and pallets of wine over the years. In a place like England, whose wine industry is, overall, benefiting from climate change, there are also a number of very real challenges. Despite the fact that the growing English sparkling-wine industry is often spoken of as being largely predicated on a warming earth, the actual situation is far more complicated than that. Even as average temperatures get warmer, the weather's variability from year to year is substantial. This is England, after all, not Napa Valley, and rain and other nasty North Atlantic weather is still a defining characteristic there. The year 2021, for example, "was really challenging," explained Trevor Clough, cofounder of Digby Fine English, a top boutique sparkling-wine producer in Arundel, West Sussex, around sixty miles south of London, right by the country's southern coast. "It's actually been really quite warm all year, and flowering was all right. It's just that we haven't had any sun all year." Excessive cloud cover isn't all that unusual for England, but the extremity of it in 2021 meant that many vineyards around that part of the country struggled to adequately ripen their fruit, even with the heat. It's a classic example of how the widespread focus on the warming effects of climate change fails to adequately take into account the full range of how it plays out on the ground and how difficult it can be for wine producers to plan ahead as a result. In 2018, Clough recalled a ripe, warm year, and the wines produced throughout England tended to be both excellent and plentiful, he said, concurring with Sugrue's assessment. "And then this year [2021], we've had the worst dud since 2012," a vintage that was utterly disastrous for grape growers and winemakers in England. "It's not as bad as 2012; it hasn't been a full dud, but it's been a really tough year."

Fortunately, Digby's model of production is set up to allow it to ride out the more challenging vintages and make the most of the good ones. They're what's known in the wine business as a *négociant*, which means that they have a portfolio of vineyard partners from which they purchase their fruit, in addition to owning their top Pinot-producing site, Hilden Vineyard, in Kent. It's an old model that for a long time has been the dominant one in Champagne. Even today, many of the most famous Champagne houses, while they generally own some of their own land, often purchase the majority of their fruit from contracted growers. The next time you hold a bottle of Champagne, look for a minuscule series of numbers on the label that's typically preceded by either the letters NM or RM (there are other two-letter combinations, but NM and RM are the most commonly found on the market). Producers from Moët & Chandon and Veuve Clicquot to Bollinger, Taittinger, and beyond are designated NM, or *négociant manipulant*, as opposed to *récoltant manipulant*, the RM designation for producers who produce Champagne exclusively from their own fruit. NM, on the other hand, means that they purchase most of their fruit. This not only allows them to produce their wine in the volume that's necessary to support the worldwide demand for their various bottlings, but it also allows the *chef de cave*, or cellar master, to have as broad a palette of base wines to work with as possible in order to blend a flagship brut NV bottling consistently each year, despite the fact that the main vintage in that blend will have been shaped by weather and climatic conditions that are unique to that particular year. (There are smaller NMs as well.) It's also a hedge against the worst years. Even catastrophic vintages typically aren't fatal in this system because though there may certainly be less fruit available, *négociants* don't just rely on their estate vineyards (if they own any) for all of their grapes; they have options. In a region like Champagne, which has historically had challenging weather (it's only a couple of hours outside Paris, not exactly the kind of place that tourists visit for its shimmering sunshine and toasty temperatures any more than they do London), the *négociant* system was a way for the producers to protect themselves from bad vintages and for the grape growers to ensure that they had customers for their fruit no matter what happened that particular year. (It wasn't always an easy or fair system, however: There were serious tensions in this relationship between growers and producers, even leading to riots in the early 1900s, but that's a different story for a different time. Today, the situation is far more amicable.)

In England, this system has similar benefits, though interestingly it's not the dominant one; most sparkling-wine producers there still grow their own grapes, which is riskier than spreading out the potential sources of fruit each year. Before England became a center of highly respected sparkling wine, the vagaries of vintage didn't matter quite as much—not only were fewer people making a living as growers of grapes and producers of wine, but the varieties themselves were different. While today it's the classic grapes of Champagne that are garnering the most attention and generating the most excitement in England, hybrids, for a long time, as Bella Spurrier pointed out, were far more common.

Genetic mutation is responsible for some of the most familiar grapes in the world, as in the case of, say, the Pinot family, whose various iterations (Pinot Noir, Pinot Gris, Pinot Blanc, and more) result from the genetic mutability of the variety. Spontaneous crossings that occurred in nature (Cabernet Sauvignon is the result of just such a crossing, in this case between Sauvignon Blanc and Cabernet Franc) account for many others. But hybrids are grape varieties that were bred for the purpose of leveraging specific chosen characteristics of the two parent plants, as was already discussed. That's why, in England, hybrids like Bacchus (from the German variety Müller-Thurgau and the offspring of a cross between Riesling and Silvaner) and Dornfelder (bred from Helfensteiner and Heroldrebe) have historically been so important; each of them possesses a unique ability to withstand the famously cool, wet weather of England. Even today, Bacchus retains a loyal following among the English wine-drinking public, Clough told me. But somewhere along the line, he added, growers and producers realized that they had the opportunity to work with grape varieties that simply produce better wine. "People said, 'Well, we've got chalk and cold weather. What about Chardonnay, Pinot Noir, Pinot Meunier?'"

The timing of England's wider adoption of those three grape varieties could not have been better. "Over the last twenty years in Sussex," Clough pointed out, "things are on average one degree [Celsius] warmer. And that has made a huge difference from those Germanic varieties [and hybrids], those really not-very-interesting varieties [from which] you can make wine, but you cannot make world-class sparkling wine." Now, he went on, with Pinot and Chardonnay on the rise, he and his colleagues and competitors can legitimately start "seeking to make world-class sparkling." And while the data set, he is quick to point out, isn't nearly large enough for him to be able to make firm conclusions about climate change in his part of England, what he has seen on the ground, in the vineyard, has proven just how much of a difference that temperature increase has made.

And though there are plenty of producers who rely on growing their own grapes, Clough is firm in his commitment to remaining a *négociant*, which he believes will allow him to both ride out the bad years with less risk and to express some ineffable truth about the character of English sparkling wine in general. "Our ambition is to make some of the top-quality sparkling wine in the world, which represents what is unique and beautiful about English sparkling wine in particular," he said. "We're really aiming for world-class quality and kind of achieving our angle of typicity within the English category." Champagne, of course, has been producing sparkling wine for hundreds of years, and consumers all over the world know exactly what to expect of Champagne in general (as opposed to other sparklers that are produced using the traditional method). Consumers also know what to expect of specific houses. Year after year, the brut NV expressions of the major *maisons* taste more or less like they did the year before, and just like they will the year after, with a few, albeit minor, variations: Veuve Clicquot Yellow Label will be fruitier; Bollinger, rounder and more biscuity; Krug, a bit nuttier; and so on. What sets English sparkling wines apart when smelled and tasted and paired with food is still being explored and discovered, though it seems as if a vivid sense of acidity and minerality will likely be calling cards. As "England's first *négociant*," Clough explained, "I work with vineyards across Kent, Sussex, Hampshire," and beyond. "I've got about half a dozen vineyards that I work with, and different soil types, different elevations, what have you. And the reason that we do that is because our mentors in Napa, when we kind of very first had the idea of building our business, said, 'You want to be a luxury brand. You want to be exporting wine. That's all nice, but a pretty label is not going to cut it. The wine has got to deliver, and to deliver, it's got to be balanced, elegant, have a lot of complexity.' And they stressed, 'It's got to have . . . style, a point of view, and good luck on the northern fringes of the possible in doing that with one vineyard year after year.' So blending is the answer, hence why we are a *négociant* and not a farmer," aside from with Hilden Vineyard in Kent.

This has resulted in wines that are fairly unique within the category of English bubbly. Because while Digby of course produces vintage expressions in the best years, it hasn't built its business or its reputation on them. In that regard, it's more focused on exploring what an English sparkling wine might smell and taste and age like from its part of the country in general, as opposed to what an English sparkling wine might smell and taste and age like from one particular vineyard and in one particular vintage. It also ages its wines for a longer period of time than most other producers, which doesn't only bring out a different character in the wines by the time its bottles are uncorked and poured, but also acts as yet another buffer against the vagaries of vintage. The 2021 harvest, for example, was, among Digby's vineyard partners, down by around 75 percent from the year before, Clough lamented. But because of its stores of reserve wine and because it sources from so many different vineyards, it didn't take the same kind of hit that it otherwise would have if all of its wines were built on a base of its own estate-grown fruit. And because Digby ages its wines for a longer period of time than most other producers, it doesn't feel the effects of a bad harvest right away and can try to plan for ways to mitigate those impacts by the time the wines are released. "Because I tend to sell quite long-aged wines . . . I've got the buffer of time," he told me. "Whereas you look at a more kind of long-established, kind of more mid-market player, which you can also find to a certain extent in the U.S., they're kind of making wine and selling it." Their turnaround time from harvest to market is shorter, and therefore, according to Clough, a bit riskier, despite the potential to realize profits more quickly. "So it's part of why my view of the industry, and Digby's strategy, is to focus on luxury," on longer-aged wines, and on hedging bets by buying fruit.

For Clough, luxury, in a very real sense, is a strategy in and of itself—not just from a brand-positioning standpoint but also from the perspective of smart economics given the challenges that are so unique to a young fine-wine industry like England's that also is coming into its own in a time of seriously unpredictable weather that's being magnified by climate change. Luxury, Clough bets, is also smart business.

As England finds its footing in this exciting and often challenging new world—and under the intense scrutiny of critics and other wine professionals both domestically and in its growing roster of export markets—its top winemakers are also trying to navigate their way into the future and what that might look like. To do that, producers like Clough are looking to the past to try to understand what they've gotten right and perhaps more importantly, what they've gotten wrong. Because while the so-called good years are getting better and better, the perception and definition of what constitutes them are changing. The bar, in other words, is being raised all the time, which surprisingly brings with it a set of its own issues.

"This is where it's not about data," Clough told me. Because for all of the growing range of metrics that can be employed both in the vineyard and in the winery and the increasing specificity of the farming and vinification techniques that are used, so much about the world of wine is still about the ancient art of the gut feeling. "You just ask yourself the questions: If I had been going for an additional ten years, if I had started, instead of 2009, in 1999, would I have had as many vintage declarations, knowing what I know now?" In other words, during those earlier times at Digby, Clough decided to produce vintage wines in some years and not others based on his perceptions of what constituted a great one at the time. But as the climate has changed—and quite rapidly, at that—the criteria in his mind of what makes for a vintage-worthy year have been defined upward: the bad years may be getting worse, but the great ones are achieving levels that would have been unimaginable two decades ago.

Interestingly, it took an epically bad year like 2012 to convince Clough that non-vintage sparkling wines would be something he'd specialize in, which as a result took the pressure off moving forward to declare vintage-worthy ones. "I didn't even think that way until 2012 informed me that I would start making non-vintage wines," he continued. "But had I, I think I would have declared less vintages in that decade than I have in the decade or so since." As a counterpoint to the difficult conditions of 2012, he looks at the generosity and abundance of 2018: "It was so warm and so lush, and the way that everyone in the industry reacted to it was, 'Wow, we have never seen anything like this.' And it's not like Dermot Sugrue [who makes Digby's sparkling wines,] had seen a year like that before. . . . It was unprecedented. It was unprecedentedly warm and amazing. And so we had really high yield for this country. And also, it was quite an amazing year for still wines," he added. It's possible, Clough believes, that 2018 will go down as one of those years where everything changed for the English wine industry, not just because of the nature of the weather and the character of the wines, but also because of what it taught the people involved in growing and crafting them: that the potential is there, given the realities of climate change and how its impacts are being felt in England, to produce sparkling wines that can easily be considered among the best in the world.

Still, though the quality of the wines is constantly on the rise, pigeonholing them stylistically is inherently speculative. Because like I've said, unlike Champagne, where sparkling wine has been produced for centuries and the character of the wines is well established,

England doesn't have anything approaching that. Its winemakers are currently in the process of sussing it out. "I think we've got decades of top-quality sparkling winemaking ahead of us, and if it gets a bit smoother and more consistent between the years, I'm happy with that," Clough told me.

As that process of growing top-quality fruit and making excellent sparkling wine continues to evolve and mature, he and his counterparts will be able to dial in what ultimately could become an English expression of sparkling wines that is widely recognized around the world. Because even in a place like Champagne, where styles are widely divergent between houses, an overall "Champagne character" is easily identifiable. In England, the best producers are looking at their land, the climate, and the weather to home in on that.

"[The wine of] England is for two things—energy and quiet sophistication," Clough said. "The energy is the primary characteristic of our terroir. . . . Growing Chardonnay on chalk [soils] in England, with our weather, all of that kind of comes together [to] give us the profile of my vintage reserve brut." In extraordinary years when the conditions provide "the kind of fruit quality and the structure to get a balance between minerality, lees aging," and the potential to age for "twenty years or more . . . that's what I declare a vintage. A lot of that is to do with our weather." Clough has embraced the changeability of the climate and set up his entire business to not just account for it but to leverage it. Across the board of what he and Sugrue produce at Digby, "I've got Chardonnay led and I've got Pinot Noir led, and I've got vintage and I've got non-vintage," though as he said, non-vintage dominates his production. "And I've got long-aged wines, blanc de blancs, and blanc de noirs," each of which shines in different years, depending on the quality and character of the fruit, which itself is directly tied to the weather. "So for a very small producer only making 10,000 cases, I have quite a lot of styles. That's just because of every year being so different, but also it being so interesting what these different soil types and these different years give me." And through them all, he is reaching for an understanding of what it means to make English-style sparkling wine.

Today, more than twenty years after starting Digby, Clough sees two distinct chapters that have informed and affected his wines. "I think in the first half of my winemaking career at Digby, as the head blender, I think it was particularly about Chardonnay—piercing acidity and linear wines. And that is the kind of hallmark of Digby. That is the core of all of my wines . . . the energy of English terroir, as opposed to [some larger producers who are] kind of going against that to try to make more round, smoother, easier-drinking wines because they're making a million bottles a year. And again, good that people are doing different things. . . . That's all really healthy." He added: "The second half of my years [at Digby], there has been more roundness and more emphasis on the Pinot Noir– and Pinot Meunier–led wines. So it does actually feel like I have two chapters, even within the very short time that I've been doing this since 2009, and that the warming has an impact."

So while he remains devoted to the crisp, energetic style of wine that has made Digby's reputation and that he believes is the most accurate representation of his beloved English terroir right now, he is fully aware of the rapid shifts being brought on by climate change. "I think that I will be able to continue making wines with our very kind of fresh and driven [character], and kind of remaining crisp and remaining fruit-forward even after years of lees aging. I think I will be able to continue making that style of wine for as long as I am doing this. But," he allowed, "I do think that as it continues to get warm and we have these maybe five-year chapters, there will be more ability to make non-vintage styles that are broadly appealing and more relaxed and more kind of safe." Those wines may begin to play a greater role in his portfolio too. Still, "England hasn't been so much for safe" when it comes to its homegrown wine industry, he admitted with pride. "We've been kind of at the bleeding edge of cool and really fascinating winemaking. But I think the category's also growing and more and more vineyards are planting. More and more people are kind of coming into the category." The famous Champagne house Taittinger now has an English fizz estate in Kent, Domaine Evremond, and plans to release the first bottles in the next few years. Taittinger's investment in English sparkling wine is reminiscent of their founding of California's Domaine Carneros in 1987. Their foresight there was spot on, and the same success is likely for them in England now, too. In the years to come, more producers are apt to pop up, and investment is likely to grow.

"So that will sort of build it out from really innovative and 'out there'" wines, Clough predicts, to more familiar, easy-to-love sparklers that boast the kind of effusive fruit and softer textures that will appeal to a wider consumer base. "There will still be those houses like me, really leaning into that [acidity and energy] and kind of saying, 'I don't need to please everyone and be just kind of bland.' But there will also be bigger bodies of bigger producers" that will make wines targeted to a larger audience. "And my goal is to be a jewel in the crown of the category and not try to be big. . . . But I think there will be more kind of filling out of the category, which is just a good thing."

As is the case with emerging wine regions around the world—or rather, wine regions that are just starting to get widely recognized around the world—it's not necessarily the boutique producers who are making more idiosyncratic wines that serve as a point of entry for the majority of consumers but rather the ones who make the kind of wines that aren't too challenging, that essentially convey the message that, yes, this is a region or a country whose wines I can really see integrating into my wine life. With Australia, for example, most consumers in the United States came to it through the world of fruit-forward, everyday-priced Shiraz from large brands, not the more challenging and complex Shiraz from Barossa's top producers, or the age-worthy Cabs of Margaret River or the energetic, savory Grenache-based reds of McLaren Vale. Those were generally embraced later. The same is true of Argentina, with value-priced Malbec having led the way. The new generation of softer, rounder, less-linear English sparkling wines that are increasingly available in the United States will likely be the ones that serve as the point of entry to that world, though of course not at the expense of wines like Digby, Bride Valley, and Balfour Winery. "We're still so small, and we're still going through a huge journey in this nation of helping the British to fall in love with their own wine. And then that spills over into the export market," Clough said. "Ultimately, the consumers have to take us under their wing and fall in love and build memories with our wines. And it's only once that happens, as a category or a brand, [do] you have the possibility to endure. And our goal," he told me with justifiable enthusiasm, "is very much to endure. I want to leave something behind that is a forever thing. That's the dream, is that people love Digby and continue to love

Digby for a long time." He could just as easily have been talking about English fizz in general. "But you've got to bring people into it, and you've got to have different producers doing different things. You've got to have good quality. You've got to have good value across lots of different propositions in a category to bring people in."

Across the unlikely landscapes of East Anglia and southern England, with their close or relative proximity to the North Atlantic, and the United Kingdom's history as a culture more attuned to the fine details of beer and whisky production than wine, England has not just continued to make a name for itself in the world of bubbly, but has leveraged the challenges of its location and a changing, often perplexing climate to create an industry that is forward thinking, open-minded, and, increasingly, producing notably delicious sparkling wine. It doesn't take a victory in a blind-tasting competition to prove that, but such exercises are often instructive—and sometimes predictive. And while the general character of the sparkling wines from England are quite different from those grown and vinified in Champagne, just a few hundred miles to the south, they seem to inch closer and closer in quality with each passing year. The trajectory of English sparkling wine is very much on the ascent, just like its temperatures. And for all the challenges it's causing in so many other parts of the world—and in England too—climate change is a main driver in the growth and improvement of the wine industry there. The future is bright . . . even in famously overcast England.

5

AMERICAN SPIRIT

The flood never should have happened this way. For years, whenever excessive rains arrived in this part of Kentucky, they proceeded in a more or less predictable pattern, one that's familiar to anyone who has lived or worked alongside a river or significant tributary: The storm rolls in, and the waterway eventually overspills its banks. Maybe there are dams or levees and maybe there aren't. If they're present and they work, then everything's fine. If there are none or they fail, then the floods come in, inexorably wending their way across the river's banks, and then if there's enough water, along streets and into low-lying homes and businesses.

But the flood that swept into the Castle & Key distillery in Frankfort, Kentucky, that late-summer day in 2018 failed to follow the usual rules.

The storm came on with speed and a sense of vengeance. At its peak, the system pummeled the area with up to four inches of rain in less than an hour. Businesses and homeowners in the area had, of course, dealt with floods coming off of Glenns Creek in the past—it's not a major waterway, but as a tributary of the Kentucky River, which is itself a tributary of the larger Ohio River, it had seen its share of weather-related problems. But those were generally predictable and manageable, aside from the occasional tornado, and the usual mitigation systems generally worked as they were intended to help minimize the worst outcomes. Just the month before, in July, a storm lashed Woodford County (among other parts of central Kentucky) with winds that gusted to more than seventy miles per hour and took out power lines, trees, and more.[1] But, as a result of the ferocity of the deluge on August 11, 2018, the floods came from above. Because of that, and because that sort of contingency hadn't really been considered before, the team at the not-yet-opened distillery most definitely wasn't prepared for it.

Up the hill from Castle & Key is an industrial park on part of the 200 acres of land known as Blanton Crutcher Farm. Initially, it was "one of the earliest homesteads in Kentucky," Brett Connors of Castle & Key told me. Yet part of the property, like so many other sprawling historical tracts across America, was eventually sold off and built up for more industrial use—in this case, warehousing. When that happened, according to some locals, the owners of Tierney Storage were more concerned with maximizing their investment than building it out in a conscientious way.[2] "So they want to build more and more massive storage structures up on that site," Connors continued. "They're just kind of . . . concerned with quick economic turnover, right?" Construction plans were drawn up and submitted to the relevant municipal departments. Once they were made public, there was some pushback to the build-out, but the local government seemed more focused on the jobs it would likely bring to the area. Construction moved ahead.

Connors told me that the problems began when the company failed to put in an "appropriate" catchment system, which would be needed to collect and drain the rainwater that would inevitably impact the area; located about an hour's drive east of Louisville, this is a part of the state that sees its fair share of wet weather. They also cut down a massive number of trees on the site. "Kind of the equivalent of deforesting, right?" Connors went on. "These are historic, massive, 100-plus-acre farm properties, and they're no longer growing anything." As a consequence of that, the water-management needs of the property were dramatically different from what they were when that land was being used for agriculture, as would the ways in which erosion would play a part moving forward.

That stripping of the trees and the transitioning of the land away from crops meant that much of the underlying soil lacked the internal structure that the roots of trees and other flora once provided—it's a relatively common issue seen around the world in deforested or overdeveloped areas. In Franklin County, there are municipal regulations that are meant to prevent this from happening, as there are in counties all across the country, but the developer didn't seem to follow them and the local authorities didn't appear to follow up enough. Before that year, it hadn't been much of a problem—rains came and went, soil was washed away in manageable and usually unremarkable amounts, and no one paid much attention. But this storm proved to be too much for the now-weakened land to handle.

When storm systems were forecast to come through in the past, Connors told me, the team at Castle & Key was able to protect their buildings. "We have regular flooding on Glenns Creek," he said. "Normally, what it will do is, it will overcrest our river's edge," and we will "be able to place some stonework down by the river, down by our spring house, [both of] which are our primary water sources. And it won't get over that." But on August 11th, the unimaginable happened. "All of a sudden, you have kind of a perfect storm of when we got an insane rain anomaly," he recalled. "Basically, storm cells stalled over the little town of Millville, right outside of Frankfort and up the hill on this road called Duncan. So all that water, which is pouring down Duncan [and overwhelming the branches of Glenns Creek], which is not made to handle that kind of an event, was coming off the watershed, off this hillside," and into the "low river valley. [And] the water didn't flood in from the creek because we're used to that." The team at Castle & Key was able to protect themselves from the worst damage from the river overflowing. But that wasn't the main problem. "Literally," he marveled, "you could see it coming off of our parking lot, off the hillside, the road. I mean . . . we ended up losing our boiler. It damaged our boiler; it damaged all of our electrical systems. It was probably a little over a million dollars of damage. Another example of why we think it was hillside as well is, we have a sunken garden on the property that's modeled after an English pleasure garden. And the water was so aggressive on the eastern wall . . . that it destroyed, absolutely just blew away a probably thirty-, forty-foot run of twenty-four to thirty-six inch, all hand-carved limestone, which is a pretty intense watershed moment," he recalled, likely not intending that pun.

By the time the rains had moved on, the debris-laden flood of mud and water that he believes started on the industrial site and washed down the hill and into the distillery had thoroughly reshaped what the team there thought their location was susceptible to, and threw their grand opening into question.

The property itself had been developed for bourbon production and tourism in 1887 by Colonel Edmund Haynes Taylor Jr., and the whiskey that was distilled and aged there developed a very good reputation. Prohibition, however, destroyed everything that had been built, and for the better part of the next century, the property—including the castle, the sunken garden, and the spring house—had fallen into extreme disrepair. It was eventually purchased by Wes Murry and Will Arvin, who set about rehabilitating and rebuilding it. They hired the much-buzzed-about Marianne Eaves to work as their master distiller and spared no expense in their efforts to bring the 113-acre property back to its former glory. And here they were, just over a month before they were scheduled to open to the public, dealing with a massive flood that because of climate change and the development of neighboring land, didn't even give their mitigation and protection efforts a chance to work.

Brett Connors has several titles at Castle & Key, including head blender and ambassador, but he actually prefers whiskey wizard, which is what appears on his business card. When I asked him about it, he explained to me that "[i]t's a little tongue-in-cheek . . . kind of a little rejectionist against people calling themselves master blenders with five years' experience . . . or one year's experience. I mean, I think wine and beer [professionals have] always done a lot better at trying to represent their skill sets and qualities in actual nomenclature, where I think whiskey, it's like, 'Yeah, I'm a master expert bourbon specialist.' And you're like, 'Are you?'" he said, a smile creeping onto his face. "There's no real continuity or governing body to actually make sure anyone's credentialed. So I like the term whiskey wizard."

Whatever title he chooses to go by, Connors is deeply involved in the world of American whiskey in general and Kentucky bourbon in particular. He has helped clients acquire and sell rare bottles at retail and auction, worked with distilleries on their marketing, and regularly interacts with consumers at all levels of whiskey knowledge. Yet even for a whiskey lifer like him, climate change is wreaking havoc on the world of American whiskey in ways that are perplexing at best and unprecedented at worst. Distilleries, for example, have "always been used to flooding," Connors told me, but the sheer magnitude and unpredictability of flooding, on the one hand, and drought on the other, combined with scorching hot summers some years and brutally cold winters in others, have made planning for the future particularly challenging.

When it comes to problems caused by rain, he said, the main issue is that people in the industry are "not prepared to handle two inches of rain in two hours, because no one is," much less even more powerful storms. "So it's like, can you advance your technological perspectives . . . to be able to catch up with a constantly, ever-changing field? In some ways I think about it, it's [like] you're playing a board game, and then your partner comes up with new rules every other turn. It's really hard to win when you're constantly receiving new rules because you're following the old set of rules. Your strategies [are built] around the old set of rules." And Mother Nature, in this era of increasingly assertive climate change, is changing the rules all the time.

Those old rules are having to be reconsidered across the entire breadth of the world of American whiskey production because whiskey, after all, is just as much an agricultural product as wine—it just hasn't historically gotten discussed in the same venerated terms. That, thankfully, is changing, because the process for producing high-quality American whiskey is every bit as detailed and challenging as it is for winemaking. And it is exposed to a similar set of climate-change risks.

It all starts with cereal grains, the specific types of which determine what kind of whiskey will be produced. The mash bill—the recipe of grains that form the basis of whiskey—is where it all starts. For bourbon, American law dictates that it must be composed of at least 51 percent corn; the rest is generally a combination of rye, wheat, and malted barley, though beyond the 51 percent corn requirement, it's up to the producer to decide what they want to use. Rye whiskey must be at least 51 percent rye. Wheated bourbons, which are still at least 51 percent corn, contain a larger percentage of wheat than is standard in the mash bill, which lends the finished product a creamier texture and a magnified sense of sweetness that, for example, is not found in the spicier bourbons that rely on higher percentages of rye. The production of Tennessee whiskey—sometimes but certainly not always spelled without an *e*, in the style of Scotch or Japanese *whisky*—is guided by almost exactly the same laws as bourbon, with the added step of filtration through sugar-maple charcoal prior to bottling. But all American whiskey (and all whiskey in general) starts off with its recipe of grains—its mash bill.

Once the mash bill is determined, the milled, blended grains are soaked in hot water, creating a "mash"—essentially a slurry into which yeast is introduced. The yeast ferments the sugars in this liquid, producing "distiller's beer." This intermediate product, which tastes like strong beer (around 7 to 10 percent alcohol by volume is the standard), is unexpectedly sweet, both because the yeast doesn't ferment 100 percent of the available sugars and it lacks the balancing bitterness of hops that are such an integral part of the production of the beers that are sold all over the world. From there, the distiller's beer is distilled in either a copper pot still or a column still, depending on the process that that particular distillery chooses. Copper pots allow for greater control of the process and often result in a more nuanced spirit, whereas column stills allow for greater consistency as well as the ability to process substantially greater volumes. Both have advantages and drawbacks, but in the end, over the course of multiple runs of the liquid through the still, each time resulting in a stronger and purer spirit, both result in a clear liquid affectionately called "white dog." Although it doesn't look any different from vodka at this point, this white dog is actually a deeply important determining factor in the quality and character of the whiskey that it will ultimately become.

During the process of distillation, three main categories of resulting liquid are produced—the heads, the hearts, and the tails. The head, the first to emerge, has a higher percentage of methanol, ethyl acetate, and aldehydes in it, which the distiller generally wants to avoid as much as possible. According to an article by Matt Strickland on distiller.com, "The overall aroma of these chemicals is kind of solvent-like and not very pleasant. Naturally, it isn't something we want to have much of in our spirit. Besides, high concentrations of some of these compounds (lookin' at you, methanol) are toxic to humans, so getting rid of as much as we can is a good practice."[3] However, Strickland adds, "A little bit of heads can create some complexity in the spirit down the line so it's up to the distiller to decide how much to cut out." The tail is laden with fusel oil and a lower-alcohol spirit, both of which are also undesirable.[4] The heart, though, represents the sweet spot, possessing a purity that both the head and tail lack and an ability to express the character of the grains in the mash bill with a notable clarity and sense of accuracy. It's also the most delicious, and the distiller has to decide when to make the "cuts" during each run of the stills he or she is monitoring. Strickland points out that, "The amount of heads and tails allowed to bleed into the heart is one of the ways a distiller decides the distillery's house character. Some distillers make these decisions based on parameters such as time and ABV. Others prefer to use taste and smell to make cuts. Either way it's as much an art as it is a science. It can take years for a distiller to become consistent with their technique." There is a constant tension between using as much of the spirit coming off of the still as possible, for financial reasons, and making sure that only the best liquid ever makes it into the barrels that will house it for the next several years.

At this point, the spirit itself is placed into barrels. Bourbon, it should be noted, cannot hit the barrel at higher than 125 proof and can only be aged in charred new oak, whereas other types of whiskey (rye, for example) don't have any barrel-specific regulations that must be followed. Yet even at this point, the decision making isn't done. There are various age requirements that determine when a whiskey can add a prestigious and highly regulated term to its label. Kentucky straight bourbon whiskey, for example, has to have aged for two or more years, and if it's been aged for less than four years, the specific length of time is required to be noted on the label. Once each barrel hits its required age, the whiskey inside can either be blended, bottled as a single-barrel expression, or finished for a short period of time in other types of wood barrels. Angel's Envy, for example, crafts expressions that are finished in everything from barrels that previously held port, rum, and Madeira. It even has one that's finished in relatively rare and expensive *mizunara* oak from Japan, an increasingly popular option throughout the world of whiskey.

At its best, all of this work is undertaken to create a spirit that is consistent year after year—a bottle of Jack Daniel's or Maker's Mark 46 or Jim Beam Black should taste the same every time you buy it. But climate change is throwing a proverbial wrench in the works and at every step along the way, challenging the received wisdom and techniques that have guided the world of American whiskey since the time of the Founding Fathers.

Most whiskey is produced using a base of commodity grains. Of course, more and more craft distillers, as well as some of the larger ones for occasional small-batch releases, are working to source locally grown cereals too; just as there has been a growing focus on the importance of supporting local and smaller-scale agriculture in the American food system, so, too, has the world of whiskey been impacted by this. But the sheer volume of raw grain that whiskey requires means that commodity cereals are still the dominant ones throughout the industry. And the climate crisis is affecting them in countless ways.

"This year's a great example," Brian Prewitt, master distiller for Sazerac, told me when we spoke in the autumn of 2021. The raw materials that he needed, the grains themselves, "Were, due to the environment, due to water constraints on some products, 20 percent short." As a result, "We're going to have to scour the Earth" to find enough of the right kinds of grains for every spirit he oversees. "I don't think that was necessarily the fact fifteen, twenty years ago. I don't think you would necessarily go, 'Oh well, let's look to all the corners of the Earth to see if we can find the product that we're looking for because maybe there is a better situation in a different country than there is here domestically.'" This, he said, poses problems for large and small producers alike. Indeed, even industry behemoths like Sazerac, under whose corporate umbrella reside Buffalo Trace, Pappy Van Winkle, Weller, Stagg, Blanton's, Eagle Rare, and more, the impacts are increasingly frightening. Prewitt may have a massive amount of money behind him to source his raw materials more broadly, but the sheer volume of spirit that he's responsible for means that he needs a stunning amount of grain too.

There are certain actions that brands and distilleries can take to mitigate their exposure to the most immediate vagaries of climate change and its impacts on the market, but many of them involve decisions that will potentially change the finished product and alter the perception of the liquid in the bottle that consumers ultimately purchase.

Contrary to popular perception, many whiskeys aren't actually crafted from a spirit that was distilled by the brand on the label. Midwest Grain Products, MGP for short, is a massive industrial distilling operation in Lawrenceburg, Indiana, that provides the spirit for countless brands; they offer customized spirits as well as "bulk barrel buys of their own maturate, [which] are incredibly common," explained Caroline Paulus, whiskey historian at Justins' House of Bourbon. "They do have a custom distillation option, but for the past ten years they've been better known for bulk sales." By necessity, many large operations rely predominantly on industrially grown grains, which are more protected from the most deleterious effects of climate change—but they're far from immune.

New seed stocks that boast resistance to too much rain or not enough, to specific insects, to various kinds of mold—you name it—make farming them a less-risky proposition. (The long-term environmental impacts are another story entirely.) But even in the case of agriculture at the commodity scale, climate change is leading to price increases that have to be passed on at every step along the production path. A lot of brands, Prewitt told me, are "adding fifty cents, a dollar per proof gallon that they're charging. And then all of a sudden, you may not necessarily notice the twenty cents or fifty cents per bottle increase, but [for the brand,] that's significant over the cost of millions of bottles of this being produced." The largest brands can try to absorb that to a certain extent, but at some point, they'll have to pass that along to the end consumer. And Sazerac, because it uses non-GMO corn in its whiskeys, faces a more challenging situation. "We're a little more susceptible to price influx here and there," Prewitt explained. "So we have to be careful with our contracts, and we have to be careful with our growing areas, because obviously [if] we lose a large portion of our crop and we're not able to find what we need for a year, that becomes difficult." When the harvest of corn from a particular area lags or is adversely impacted by that year's weather, "You have to make these decisions like, 'What are we going to do? Are we going to pay more money? Are we going to make less product? What's the choice?'"

For smaller whiskey producers, the choice is often one of financial life and death: If they produce less whiskey, then they have to raise prices to make up for that loss. And for all the staggering growth of craft whiskey in America, there will inevitably come a point when even the most exciting bottles will be priced out of reach of enough consumers to become a weight on the bottom lines of the brands producing them.

Then there is the issue of new brands and their ability to simply get a whiskey label off the ground. Whiskey, after all, takes time to make by its very definition. For the best of them, and in the most heavily regulated categories, there is no replacing the most time-consuming aspect of production, aging. When I was writing this chapter in the second half of 2021, a search for Castle & Key bourbon online directed me to a page with a photo of a barrel and this message: "Castle & Key Bourbon: Aging Until It's Ready." Below that was a button that took me to a list of the spirits it *was* selling in the meantime, including multiple gins, two whiskeys, and a vodka . . . but no proper Bourbon. By March of 2022, when I was just finishing up editing the book, Castle & Key was finally ready to launch their much-awaited bourbon. For even smaller, less-well-capitalized brands, the barrier to entry is incredibly high, which forces even whiskey-focused ones to produce other spirits to keep the proverbial lights on in the beginning. Because even if they're purchasing their unaged spirit from a large third-party distiller—a significant expense, especially at volume—they still have to wait for that liquid to repose in wood before it transmogrifies into a high-quality whiskey. How, then, can they make ends meet during that time? Fortunately, as Paulus noted earlier, MGP, for example, also offers a huge range of aged whiskeys, which brands can then purchase and blend on their own.

For many new producers, they look to both vodka and gin, which don't need to rest in barrels because they're clear spirits by definition. (Barrel-aged gins are a deliciously unique category, but they are a small part of the gin industry.) There are a number of advantages to producing gin, chief among which is the ability to turn out product relatively quickly, especially in comparison to the time required for whiskey. It's also a good way for the new producer to earn its bona fides, to build a reputation among the customers who will ostensibly eventually become their core group of supporters once whiskey is being sold.

One of the many appealing aspects of gin is its ability to express the specific place of origin of its constituent botanicals. The Botanist is a good example; it's crafted at the Bruichladdich distillery on the isle of Islay, Scotland, and incorporates herbs, flowers, and fruits that are unique to that part of the country, including bog myrtle, gorse, heather, and more, in addition to other components that are brought in from other parts of the world. But there are plenty of gins being produced around the world and across the United States that are less about the expression of the botanicals of that particular patch of the planet and instead are based more on the vision of the distiller or blender him- or herself. Revivalist Gin, for example, releases five different gins each year from its home base in the Brandywine River Valley of Pennsylvania. Each of them is an evocation of the season—the autumnal Harvest expression boasts the warming spices so closely tied to fall, whereas the Summer Equinox speaks of flowers and sun-warmed hay, and so on.

But even gin, which is supposed to make life easier for distilleries looking to generate a cash flow that's faster and less contingent on waiting for time to work its magic on a spirit in a barrel, is being challenged in unexpected ways by the changing climate. "It's been tough," lamented Prewitt, who, in addition to overseeing whiskey and brandy production for Sazerac, is also responsible for its gins, including Tinkerman's, Booth's, and Miles'. "A couple years ago, heather was really hard to get. And then last year, because of the weather again, citrus-producing countries [were unreliable]. . . . All these people that use citrus peels and citrus extracts for their gins, it was impossible to find citrus. There's a lot of pest things going on that are behind that as well." Bergamot, for example, which lends a woodsy, orange-like character to many gins (as well as to Earl Grey tea), hit $80 per pound, Prewitt told me—a new high. "It's ridiculous!" He also incorporates natural vanilla into one of the gins he works with, and it recently rose to $1,000 per pound. And while he doesn't need to use a lot of it to benefit from its flavor and aroma, that's a significant amount of money to invest in vanilla, especially considering the scale at which Sazerac produces.

The increasingly unpredictable changes in the climate are pinching production and forcing prices to climb in equal measure. "You have the increased tropical storms wiping out the crop, the cocoa and the vanilla crop . . . that's part of the driver," he went on. "But I also think it's the changing environment that's driving some of these things. Different weather patterns that are just driving a lot of those, just the core pieces that people don't really consider. When you think about spirits, you wouldn't necessarily think about [these other components]. You may think about corn, you may think about wheat. But you may not necessarily think about, 'Hey, what about all those botanicals in gin,' right? All the juniper berries, or what about making a grape spirit when you're talking about brandy? I think all those areas are really getting impacted significantly," and producers large and small are feeling it.

He added, "If you're a new craft distiller and you want to ultimately release whiskeys, I mean, you've got to turn a profit at some point and have to age in wood. So they're making gins, right? They're making these clear spirits; it's easy. But now, all of a sudden, that almost seems like that's going to be priced out of reasonableness at some point."

An increasing focus on local botanicals "definitely helps mitigate some of the factors" of the supply chain, he said, but that can't solve all of the problems brought about as a result of climate change. "It's as if a baker goes to his shelf, and he goes to look for not your standard bread flour but the different kinds of flour that he uses for pastries or whatever, and [he] just can't get it." Those key components, from the grains that a whiskey is distilled from to the botanicals that give their inimitable character to a gin—all of them are becoming much fickler than they were in the past. And at some point, even the quick cash flow option of gin won't be easily justifiable for many but the most heavily capitalized new brands. What will happen to the world of American spirits then?

There are options, however, to speed up the process of aging whiskey, which as I've said is one of the main financial pinch points in getting a new whiskey brand off the ground. Just like a vintner won't get grapes of a high enough quality to make great wine from a newly planted vineyard (even the most conscientiously planted vines in the best terroir won't produce palatable wine for three to four years) so, too, is a spirit not likely to become a high-quality whiskey with less than two to four years of aging (though there are exceptions); all of the color and much of the character of the finished whiskey comes from the time it spends resting in wood barrels unless we're talking about flavored and color-manipulated whiskeys, which are an entirely different thing altogether. Though great whiskey is always made from a high-quality mash that's distilled with care and an obsessive attention to detail—you just cannot make great whiskey from a low-quality spirit—you also can't make great whiskey without allowing it to rest for the appropriate length of time in wood barrels. It's a deeply symbiotic relationship.

The process is fairly straightforward. A barrel is chosen based on the source of its wood, the cooperage that crafted it, the level of toast or char that has been applied to the interior, its size, and whether it's new or used (and if it's used, what was in it beforehand). Once the barrel is chosen, it's filled and then allowed to rest for a period that in some rare cases can last for decades. However long the maturate spends in the barrel, it slowly and inexorably pulls a wide range of compounds from the wood itself as it rests inside. Everything from tannins to vanillin to wood sugars to lactones (which lend those classic coconut-tasting notes) and more are slowly incorporated into the liquid from the oak, changing its character and turning it into whiskey as we know it. But this process doesn't happen at a constant speed and at an even progression; depending on where in an aging warehouse a barrel is located, exposure to heat and humidity and temperature swings will affect it. Barrels that are stored higher up in the racking are exposed to more heat than ones stored closer to ground level, especially if the warehouse isn't temperature-controlled. The whiskeys in those barrels tend to be more oak-influenced than the ones that experienced less heat. Even in temperature-controlled warehouses, the height at which a row of barrels rests will affect how much heat it's exposed to; a seemingly insignificant difference of a single degree Fahrenheit, magnified over the course of four years of aging, will leave its unique, undeniable mark on the liquid.

Over the generations, producers have found ways to leverage this, blending barrels from various locations throughout a warehouse to create a final whiskey of greater complexity and harmony than any single barrel is generally capable of. Other producers rotate the location of their barrels to more evenly expose each one to the range of conditions in their warehouses, though that is a time-consuming, cumbersome, and costly process. Single-barrel expressions generally leverage the unique character of one particular barrel that has imbued it with a personality that's idiosyncratic enough to warrant being sold on its own. (In the case of the Four Roses 100 proof single barrel expression, the producer consciously looks for certain barrels that have a particular set of aromatic characteristics and bottle it as such, meaning that its single-barrel releases are remarkably consistent, which isn't often the case with other producers. Depending on who you ask, that variability is either part of a single-barrel whiskey's charm or its Achilles' heel.)

But all of this takes time. The wood expands in the heat and contracts in the cold, and throughout this "breathing" over the course of the years, the liquid in the barrel penetrates into the wood, pulling more character from the individual staves, and then it's essentially squeezed out as cooler temperatures contract it all. The barrel, essentially, respirates whiskey. The longer it does so, the more flavors and aromas the whiskey pulls from the barrel.

Unfortunately, there is a very real cost to this, and not just in time. The longer the liquid spends in the barrel, the more of it will evaporate. Depending on where a warehouse is located, whether it's temperature-controlled, and the conditions the barrels inside it are subjected to, this so-called angel's share can account for two percent or more of the liquid in the barrel every year—a major source of lost revenue. This is one of the reasons that older whiskeys cost so much more; their rarity isn't just a product of fewer of them being made but also of how much of the liquid itself is lost to evaporation through the barrel with each passing year. And as climate change makes both hot and cold weather swings more extreme, whiskey producers will have to either accept losing more from angel's-share evaporation or invest in more efficient temperature-control systems for their warehouses, which is both expensive and potentially, in the long term, detrimental to the environment. (Re-purposing the heat generated from other parts of whiskey production is one good option, however.)

Lately, however, a number of companies have introduced technology that shortens the amount of time a whiskey needs to spend in a barrel to gain the character that previously could only have been achieved by the passage of time. Experiments have been ongoing that are measuring the impact of using smaller barrels, which changes the ratio of liquid to wood, or exposing barrels to short periods of higher heat or to artificially extreme temperature swings to speed up the interaction of the spirit and the wood.[5] The results have been compelling, but heating requires additional energy, which doesn't seem to be a great option given the realities of climate change. It also still requires the same amount of wood to be used for the barrels, which is cumbersome, expensive, and not terribly sustainable; trees are a finite resource, and even replanting them requires waiting decades or more before they've grown enough to be harvested again.

The technology behind Bespoken Spirits, then, is intriguing. The company was launched by Stu Aaron and Martin Janousek in 2018. By 2020, they had already attracted the attention of investors that included, in one tranche, a group in which ex-Yankee star Derek Jeter was involved. The crux of it, according to Aaron, is fairly straightforward: "We've invented a process that's all about mastering the maturation of aged spirits," he told me during a Zoom conversation and tasting. "We call that process craft maturation, and it's about helping to take craft and creativity to whole new heights by blending both technology and tradition, and by applying the same level of art and science that has historically gone into fermentation and distillation, but not maturation, finally to the maturation process as well." Which, if I'm being honest, sounded at the time like an awful lot of well-rehearsed boilerplate when he first rattled it off, a pitch to go along with a PowerPoint presentation.

But here's the thing—the spirits are solid, and as he and Janousek took me through the technology powering it all, I couldn't help but think that in a world of ever-more-limited resources and increasingly unpredictable climate, Bespoken could actually be an important player in the future.

Before Bespoken, Aaron was involved most notably with the video-conferencing company BlueJeans; Mixpanel, the data-analytics company; and Bloom Energy, which focuses on green energy. Janousek worked on the technology development and engineering side at Bloom Energy. "Martin and I are actually a couple of serial Silicon Valley entrepreneurs," Aaron went on. "We actually first met each other fifteen years ago, when we worked together and launched another sustainably focused company [Bloom Energy], which was a fuel-cell company, a fuel-cell startup. . . . And at Bloom, Martin ran the hardcore technology development, and I ran the marketing and the business development function, and you can probably guess by that high-level description that Martin is a materials scientist by training, [but] he's a foodie and a beverage aficionado by passion. Whereas I'm admittedly neither of those things. I'm the business guy, and I've got more of the everyman's palate."

He continued: "So Martin approached me with the idea behind Bespoken first about three and a half years ago. What happened was, he was getting really frustrated in his wine and whiskey club, that in order to get an awesome bottle of something, he either needed to spend a lot of money, [or] buy the bottle years in advance and just sit on it, and he didn't understand why it had to be that way. And so, like any good scientist, he started reading and tinkering and inventing, and he came up with the technology behind the company."

It's that technology that has the potential to make a real impact. Their timing couldn't be better. The gist of it is this: "Rather than putting the spirit in the barrel, and waiting hopefully but helplessly and passively for nature to take its course, we instead use our proprietary ACTivation technology to instill the barrel into the spirit," Aaron explained. "By carefully sourcing and characterizing and controlling the wood that we craft and blend together for our process, as well as the chemical reactions and the environmentals, we are able to to deliver premium quality tailored spirits and get what we want out of the process, and to do it in days rather than [years or] decades."

This is accomplished by Bespoken's deconstructing of the barrel. "Our process, 'craft maturation,' uses the same natural elements of wood, toast, and char, and source spirit that's used in traditional barrel aging. Nothing else—there's no additives, there's no chemicals, it's just spirit, wood, toast, and char. We have just reimagined the process with modern materials science and data analytics to get more control and precision . . . around that process." He elaborated: "What we do is, we begin the process by sitting down with the customer to design the target spirit that they want it to become. And what we're looking for is an exact aroma, color, and taste profile that they want. That's most analogous to the Pantone wheel for color you'd find in a paint store, except instead of defining colors like burnt orange as red, yellow, and blue combinations, we're defining smokey whiskey or French toast as aroma, color, and taste combinations. And sometimes our customers know exactly what they want: 'I want this much vanilla, I want this much caramel, I want this much citrus.' Most of the time, however, it's more of a reference. 'I want it to be like this,' and they'll point at a known brand. Or, 'I made this five years ago, I want it to be like this,' or something like that. And we use that guidance to develop recipes that deliver that result." And every time they plug in a new recipe, their system, through machine learning, has a better chance of perfecting subsequent ones.

"In developing our recipe, there's three main elements that we use. The source spirit itself, because while we can take a spirit in a wide range of directions, the source spirit does impose some ultimate limitations. In other words, if it doesn't have peat, we're not really going to bring out peat through the wood, as an example. And then the wood is the second key element of our recipe, and then the third is the settings on our machine which we call the ACTivator, where the ACT stands for Aroma, Color, and Taste. . . . Some people see our process and they think because it's short, it's easier, whereas actually it's a lot more work that we put into developing this recipe, because what we do is, we hand-craft a mixture of wood for the result that we want to achieve. We use small pieces of wood that we call microstaves. And each microstave is roughly one-25,000th the size of a barrel. So, think about it as maybe half the size of your little finger. And because it's so small, each microstave brings to our recipe exactly what we want it to bring from a wood perspective and none of what we don't want it to bring. Contrast that to a barrel, which because wood is a naturally occurring [material], in any barrel and any stave you get some of what you want, some of what you don't want, never in the right ratios, [and] never consistent. Whereas we, because we source and we characterize every microstave, we know exactly what we're getting based on which microstaves we choose for our recipe." From there, they mix the microstaves using "different types and quantities and ratios," Aaron continued. "So for one recipe, we might use all American oak; for another recipe, we might mix some American oak with some French oak with some cherry wood."

"And so," he added, "once we have our source spirit and our microstave mixture put together and toasted and charred, we put them together in our ACTivator, and within the [machine], we're controlling the environment, the temperature, the pressure, the agitation rate, the atmosphere, etc. And what you find is between the source spirit, the wood mix, and treatment, and then the ACTivator settings, we have over 20 billion unique combinations to work with in developing our recipes. And they yield different results." This process can be used to both create new spirits or to fix spirits that when they're removed from the barrel, haven't achieved the desired set of characteristics. From an environmental standpoint, it's remarkably earth- and energy-friendly—a "sustainable process," Aaron told me. "[C]ompared to traditional methods, we're typically using 97 percent less wood, 99 percent less energy, at least 20 percent less water, and we have zero angel's share to speak of." When I asked him about the energy required to run the machines and the computers behind them, he argued that "[e]ach ACTivator doesn't use a ton of energy, and the point is we're only running it for three days, compared to five years or more that you're powering a rickhouse, for power, cooling, security, lighting, etcetera. And so you do the math, and it turns out that we're a fraction of a percent of what's typically being used on the energy front."

As for the whiskeys themselves, I've been impressed, despite my initial skepticism. The best of them are detailed, balanced, and well-crafted. The question, as always, is how consumers will react to something so new and unfamiliar. "As the whiskey historian at

Justins' House of Bourbon, I help curate the largest collection of bourbon for sale in the entire world," explained Caroline Paulus. "While we have a huge range of vintage expressions, we also have hundreds of modern bottles from big names like Woodford Reserve as well as smaller craft makers. In my experience, consumers are more likely to try a whiskey whose bottle features familiar terminology, a familiar phrase like 'straight bourbon whiskey.' But until a bourbon drinker tries a product made with accelerated aging techniques and actually enjoys it, it'll be a tough sell for most people—I think they'll be less likely to take a chance on a bottle without that language on the label. What Bespoken is doing might be the next wave," she added, "but bourbon drinkers and makers have always had a healthy respect for the years it takes to craft a good whiskey. That base of experience and expectation is the hurdle they'll have to get over, but as long as consumers find that they enjoy the liquid itself, then it could find an audience."

As the technology continues to mature and become more widely accepted in the world of whiskey, it could become not just a viable tool in the belt of distillers and blenders but also a potential way forward in a world that's becoming increasingly challenging for whiskey brands both old and new. I don't see it ever replacing whiskey production as we know it—nor should it—but I do think it'll become more widely accepted as another weapon in the proverbial arsenal.

For all of the challenges that climate change is hitting the world of American whiskey with, there are a number of fundamental shifts happening, especially among smaller brands, that could ultimately prove beneficial. Among them, few may turn out to be more consequential than a greater focus on the kinds of grains that are actually meant to grow in a particular place. Most of the large producers, as I've mentioned, contract for their grains—if they're distilling their own spirits at all. Yet even among the ones that *are* doing their own distillation, "The vast majority of distilleries don't know where the grain is coming from with any specificity," explained Adam Polonski, cofounder of Lost Lantern along with Nora Ganley-Roper. Lost Lantern is an independent bottler located in Vermont, which since kicking off business in 2019, has released whiskeys from small and midsized distilleries in more than a dozen states. Their model involves Polonski and Ganley-Roper traveling the country and looking for standout barrels, which they then either have blended or bottled as is, to be labeled with both their name and the distillery's. As such, their work takes them around the United States, and they've been able to see firsthand how the world of American whiskey is being changed by shifts in the climate.

The reason that so many large-scale distilleries don't necessarily know where their grain comes from (and again, there are important exceptions; even within distilleries whose biggest expressions are crafted from commodity grains, there are more and more examples of small-batch bottlings that utilize less-anonymous cereals) is because "[c]ompared to grapes," Polonski told me, where grain is grown "doesn't make as much of a difference [at that scale]. And that's in large part because the vast majority of grain for whiskey is commodity sourced, and it's coming in bulk from Minnesota, Saskatchewan, Alberta, wherever else it may be. Whereas someplace like Romanée-Conti," the famed vineyard in Burgundy whose Pinot Noir fetches thousands of dollars per bottle, "they've been farming grapes on this one vineyard for [nearly a thousand years]." As a result, the inimitable character of that one vineyard rings through with stunning clarity vintage after vintage. Since "there's no expectation for that in whiskey," he continued, "they can shift the sources around a little bit over time" without any great impact on the character of the final liquid in the bottle. Plus, crafting widely available whiskeys relies a great deal on blending, and any small shift in flavor and aroma in the corn or wheat or rye from one place or another can be compensated for in the final blend by leveraging the massive stores of barrels that crowd the warehouses of large producers. "I think they can maintain that consistency if they choose to," Polonski told me. Interestingly, he added, the idea that the majority of a whiskey's flavor comes from the barrel is in many ways a result of commodity grains, which have generally been bred for hardiness as opposed to flavor.

Noncommodity and heritage grains, on the other hand, are very different. And while they pose their own challenges—the higher amount of protein in many heirloom wheats, for example, makes them more finicky during the fermentation process—the rewards can be substantial. Dad's Hat Rye is an interesting example. It was created in Bristol, Pennsylvania, in 2010 by Herman Mihalich and John Cooper, and from the start, their goal was to bring back the glory of Pennsylvania rye whiskey, which had a sterling reputation before Prohibition nearly wiped out the industry in the commonwealth. For them, distilling with commodity grains would have been cheaper and easier but less authentic. So they have been working with local farmers and several university agricultural departments to not just focus on grains that are well suited to the various regions of Pennsylvania that they're being grown in but that also have, in some cases, been unused for decades. From seed repositories to soil scientists, they are working as hard as they can to mine the true character of what a Pennsylvania rye can and should taste like. The results have been widely praised by critics and journalists, and consumers have been snapping it up. Still, as the climate continues to shift and become more extreme, smaller brands like Dad's Hat and the other distilleries that Lost Lantern works with will likely find their efforts challenged more than they ever have been before. And since they don't have the economies of scale that the large ones do, their prices are likely to climb far faster. The question hovering above it all is whether consumers will accept that.

Small brands also don't generally have the luxury of a massive store of barrels—certainly not compared to the large producers—and therefore are more susceptible to changes in flavor and aroma from one batch or one year to the next. Polonski has noticed that "[s]ome of the smaller distilleries now . . . are doing something more along the lines of traditional wine vintages, where they will have a release which is like, 'This is this year's harvest,'" he said. In Scotland, Port Charlotte has had success with this model, releasing their "Islay Barley" line of single malt Scotch whiskies labeled with the year the liquid was distilled, the year the whisky was bottled, and credit on their website to the specific farms where the barley was grown. Some smaller American whiskey producers are headed in that direction too, at least for some of their expressions. Even the larger ones are dipping their proverbial toe in the water of working with single-farm grains, such as Woodinville, the increasingly popular Washington State whiskey producer, which boasts a back label that reads as if it were a small, mom-and-pop operation, despite the fact that it was acquired by LVMH in 2017. "This truly small-batch spirit starts with traditionally grown corn, rye, and malted barley. All of our staple grains are cultivated exclusively for us on the Omlin Family farm in Quincy, Washington. The grains are mashed and distilled in our Woodinville distillery, then trucked back over the Cascade Mountains to our private barrel houses, where Central Washington's extreme temperature cycles promote the extraction of natural flavors from the oak." This kind of sourcing benefits both the farm *and* the corporate parent behind the whiskey, giving the former a reliable customer for their grain and the latter the ability to lock in pricing from a single farm rather than dealing with the increasingly manic fluctuations of the commodity grain market. And the glow of environmental responsibility is likely not lost on the LVMH team either. Plus, the whiskey is delicious.

Other industry behemoths are actually leveraging climate extremes when the opportunity presents itself and trying to use them to create new styles of whiskey that interface more directly with the changing environment. Buffalo Trace, which is estimated to produce more than 2.5 million gallons of whiskey each year under labels as diverse as Weller, Pappy Van Winkle, Stagg, and more, released an E.H. Taylor Jr. Warehouse C Tornado Surviving Bourbon. According to the brand's website, "On Sunday evening, April 2, 2006, a severe storm with tornado strength winds tore through Central Kentucky, damaging two Buffalo Trace Distillery aging warehouses. One of the damaged warehouses was Warehouse C, a treasured warehouse on property, built by Colonel Edmund Haynes Taylor Jr. in 1885. It sustained significant damage to its roof and north brick wall, exposing a group of aging bourbon barrels to the elements. That summer, the exposed barrels waited patiently while the roof and walls were repaired, meanwhile being exposed to the Central Kentucky climate. When these barrels were tasted years later, it was discovered that the sun, wind, and elements they had experienced created a bourbon rich in flavors that was unmatched. This was truly a special batch of barrels, and though the Distillery does not hope for another tornado, it feels lucky to have been able to release this once in a lifetime product."[6]

Talk about making lemonade from climate lemons! And it's not limited to just that one expression. Buffalo Trace has an entire line, called the Experimental Collection, that includes "more than 30,000 experimental barrels of whiskey aging in its warehouses. Each of the barrels has unique characteristics and experimental changes in the mash bill, types of wood, barrel toasts and more. Periodically, an experimental whiskey is bottled and sold on a limited basis."[7]

On a much smaller scale, some American craft distilleries are more and more frequently being forced to go through these sorts of experiments every time they run their stills and fill their barrels, especially if they're moving away from the commodity grain market and more toward the smaller farm one. Frey Ranch in Nevada, for example, grows its own grains for the whiskeys it produces. The fact that it raises its cereals on a specific patch of the planet means that it does express the vagaries of climate more intensely and clearly than the bulk-grown ones that larger producers blend away into relative flavor and aroma anonymity. The result is a more location-specific spirit but a riskier one to produce.

This brings us back to the idea that "so much of whiskey's flavor comes from the barrel," Polonski said again. "It's not because it was inherently true. It's because you're buying cheap grain that doesn't necessarily have as much flavor in itself. So then," out of necessity, "the flavor comes from the barrel. If you're using really high quality, locally grown grain in the first place, then that brings a lot more flavor overall, [but] then I think you will see more climate variation" in each harvest. Ganley-Roper predicted that "[t]here will be a subset [of smaller distilleries] that really rise up as the creative leaders when it comes to thinking about vintages, thinking about working with local farmers and what that means for the whiskey and the stories they tell. We'll see, and we're already seeing . . . a subset of distilleries that really are able to show something interesting on that front."

Plus, it's better for the land, especially if those grains are being farmed sustainably. Industrialized intensive farming, after all, with its not-uncommon chemical inputs—pesticides, herbicides, and the rest—is bad for the earth in the long term. But working with and supporting smaller-scale farmers, Polonski added, "[c]an be a hedge in a small way against some of the agricultural impacts of climate change, especially because so many distillers are working with local farmers and not having to grow commodity crops, or having them grow crops that are [beneficial] for that environment, that tend to be better for the soil [and] often help . . . regenerate the soil. And we see that across the country, where places that have traditionally been farmed to very heavily fertilized crops with pesticides and things like that, just switching over to organic. . . . You're not taking the nutrients out of the soil." A number of larger operations are making the switch, too.

As climate change picks up speed, the coming years could dramatically shift the paradigm in the world of American whiskey, from the most boutique brands to the biggest ones. In the meantime, smaller distilleries, more nimble major companies, operations like Bespoken Spirits, and independent bottlers like Lost Lantern are all leveraging today's challenges to release thoroughly unique spirits. "For us," Ganley-Roper said, "our whole business is about finding the interesting, unique things that are being done. So these variations across the country, we have an even broader range of palettes to look at, or flavors to look at, and our palette range is larger. So there is an interesting kind of growth in what we're able to work with through this not-as-good local environment."

The world of American whiskey is changing dramatically. And while consumers, especially of the major expressions of the largest brands, aren't seeing the impacts directly quite yet—aside from occasional price fluctuations or supply chain–related shortages—the entire underlying structure of the world of American whiskey is slowly starting to shift. Climate change is driving this, and from droughts in the Upper Midwest to floods from above like the one that washed into Castle & Key in 2018, business as usual will likely look very different in ten years from how it does right now. In the short term, it could ultimately benefit consumers—prices may climb, but the diversity of options could be substantially broader that they are even today. But at what cost in the long term?

6

WHAT HAPPENS IN THE SOUTH WHEN TEMPERATURES GO NORTH

As my flight from Buenos Aires to Patagonia was making its initial descent into Neuquén airport (it's officially called Aeropuerto Internacional de Neuquén—Presidente Perón, but everyone tends to refer to it as Neuquén) in 2010, I have a distinct memory of looking out the window, prepared to see a land reminiscent of the icy *Star Wars* planet Hoth, with snowdrifts and glimmering ice, maybe even the occasional blue-glowing glacier, as far as I could see. Instead, we landed in a place that looked a lot more like Montana, cut through with rivers and carpeted with rolling fields of breeze-blown wild grasses. This region, I discovered over the following few days, was full of surprises.

That was more than a decade ago, before the reality of climate change had begun to outpace even the direst predictions of scientists at the time. It was notably cooler there than in the more northerly parts of the country I'd just been visiting, but nothing like the snowy hinterlands I'd built up in my mind. No wonder so many producers had begun moving there to stake their claim: the land was plentiful, the rivers crisscrossing it meant that there was plenty of water for irrigation, and the chillier temperatures meant that cool-climate grapes would be able to thrive there. Of *course* the scion of a major Italian winemaking family had decided to try his hand in Patagonia just a few years earlier; this was a new wine frontier as full of promise as any I'd ever set foot in.

As I write this, eleven years later, the benefits of growing grapes and making wine in Patagonia are more apparent than ever . . . and in some ways, also less certain. Because no matter where you move these days, climate change is bound to find its way in.

Wine from *Vitis vinifera*, the classic grapevine species of which Cabernet Sauvignon, Chardonnay, Pinot Noir, and the other standards are part, has been made in South America for approximately 500 years. It came to the New World with the Catholic Church, wending its way through Mexico, Central America. Peru, and beyond. In the mid-1500s, when conquistadors from Spain arrived, the missionaries who came with them needed wine. Records still exist of a letter, according to a history on the website of the Chilean producer Concha y Toro, that was sent from Pedro de Valdivia to King Carlos V requesting "grapevines and wines to evangelize Chile. . . . Following the letter, a shipment of wine from Peru and the first grapevines from Europe arrived."[1] Interestingly, the history continues, "Despite the fact that the vines were planted in different parts of Latin America, only when they reached Chile did they find that perfect combination between climate and soil to produce grapes worthy of being made into wine"—or at least, wine worthy of a reputation beyond its immediate areas of production.

It makes sense: Chile and Argentina are uniquely well suited to the production of wine. Chile is tucked between the Pacific Ocean to the west and the Andes Mountains to the east, which means that its proximity to the water and to the cooling currents that are funneled northward from Antarctica, moderate temperatures. The Andes, the highest mountain range on the planet outside of the great ones of Asia—the Himalayas, the Karakoram, and more—provide the opportunity for planting vineyards at varying elevations and allow for natural irrigation through the melting of snow and its running off into the lands below. Argentina also benefits from the Andes, which over millions of years, have created a deeply complex series of valleys, each with its own unique geology and microclimate, perfect for growing high-quality and varied wine grapes. On top of that, the Pacific and the Andes have historically provided protection from certain environmental and other phenomena in Chile, most notably the phylloxera epidemic, the vine louse that virtually demolished the wine industry of Europe in the second half of the nineteenth century. Chile remained more or less phylloxera free for centuries as a result of the natural barriers of the Pacific and the Andes.

When Europe was being overrun by that notorious vine louse, in fact, Chile and Argentina benefited from the Old World's loss; many grape growers and winemakers from Europe who could no longer ply their trade at home moved to South America, bringing with them knowledge, technology, and different grape varieties than had been planted there before. Missionaries had introduced the País variety and grown it widely hundreds of years earlier, and it remained the most widely planted *Vitis vinifera* red until the nineteenth century rolled into the twentieth (at its best, País, also called Mission, can produce compelling wine, but I haven't yet tasted anything profound from it). But it was the arrival of the Italians, the Spanish, and the French in the mid- to late-nineteenth and early-twentieth centuries that really changed everything for the South American wine industry. Even today, for example, though Malbec is Argentina's most famous and successful grape variety (and the wines produced from it in and around Mendoza have done more than any others in making Argentina's wine reputation around the world), it wasn't from there originally; its ancestral home is in France. According to a piece by Tom Bruce-Gardyne, "Back in the Middle Ages, Malbec was planted all over southern France. But it wasn't known as Malbec. It had over a thousand synonyms, and besides Medieval wine drinkers knew precious little about grape varieties. But there was no doubt Malbec was highly thought of, especially up-river of Bordeaux, where it was blended with the even darker Tannat grape to make the famous 'Black Wine of Cahors.'" He added, "Malbec was one of a number of vines introduced into Argentina in 1868 by Miguel Pouget, a French agronomist who had been hired to help improve the country's wines. It seems the particular clone he brought over has since disappeared in France. Argentine Malbec has smaller grapes and tighter bunches than the type grown in Cahors."[2] In Argentina, where Malbec benefits from the often high-altitude vineyards of the eastern slopes and foothills of the Andes, it produces a wine of power on occasion but also of complexity, lift, and energy . . . wines that are often rich, nuanced, and elegant.

Wine is produced in other countries in South America—Brazil has a burgeoning industry, and Uruguay produces some notable wines too—but most of the continent's wine is produced in Chile and Argentina. It's important to stress, however, that though they are often spoken of in collective terms, Chile and Argentina are very different from one another in culture, history, cuisine, wine, and how climate change is impacting them individually.

There are, however, some key similarities, most notably the fact that the modern-day wine industries in both countries have consistently grown and improved in recent decades. New vineyard areas are constantly being discovered in already existing wine regions, affording grape growers the ability to coax novel and exciting flavors and aromas from the grapes as they channel and translate the character of the land in which their vines are planted. And markets around the world increasingly clamor for the wines produced in these two countries, from everyday-priced bottles to trophies given pride of place in cellars around the world. Critics swoon with more passion vintage after vintage; in 2014, for example, *Wine Spectator* magazine ranked the Concha y Toro Don Melchor, a Cabernet Sauvignon–based red from the high-altitude Puente Alto vineyard in Chile, its ninth-best wine of the year, an honor that had never before been given to a wine from South America. The wines of Chile and Argentina are indisputably on a tear.

But beneath the surface, like in some sort of horror movie, problems are brewing. And the biggest of them are a result of climate change.

Just as more northerly areas in the Northern Hemisphere are benefiting from the warmer temperatures brought on by climate change—English sparkling wine is a great example, as are the pioneering producers of Denmark—so, too, are the more southerly reaches of the Southern Hemisphere. In South America, the search for cooler climes has led, perhaps inevitably, to Patagonia.

Grapes have been grown and wine has been made in this massive region for over a century, but the modern wine industry as we know it didn't find its way to Patagonia until much later. For Guillermo Barzi, commercial director, and Juan Martín Vidirí, director of production of Humberto Canale winery, both of whom represent the fifth generation of their family in the wine business in Argentina, Patagonia is nothing new. Its growing popularity, however, is.

Humberto Canale was born in 1876 in Buenos Aires; his parents had come to Argentina from Genoa, Italy, and worked in the baking industry. "Remember," Barzi told me, "at that time, there were a lot of immigrants in Argentina. Most of us, we came from Spain and Italy, so we have a lot of culture of wine. Many of these relatively new arrivals to Argentina worked on surveying and construction projects in the newly accessible frontiers of northern Patagonia." And continuing the practice that was so important back home in Europe, "They used to have their own vines in their houses," small vineyards that they tended to have grapes to ferment for personal wine consumption. "So Humberto Canale," Barzi continued, "the first thing he [did] in 1909, he planted the first sticks from Cabernet Sauvignon. And so our winery was [born]." The year 2021 represented the family's 112th harvest, making it, he said with pride, "one of the oldest wineries in Argentina in the same family."

This is all to say that making wine in Patagonia is nothing new—the urgency with which producers have been moving there, however, is. For years, Barzi explained, he and his family worked in relative solitude, in the company of "maybe two or three wineries around the area." Now, however, that has all changed. "I don't know exactly the number, but we think that in Rio Negro," the region of Patagonia where Humberto Canale is located, "we are [now] around more than thirty wineries."

This increasingly widespread move to Patagonia makes sense; as temperatures have climbed farther north, there is a serious opportunity to not only avoid many of the problems associated with hotter growing conditions there but also to work with the kind of grapes that are becoming increasingly appealing in the cooler climates of Patagonia. Cabernet Sauvignon is a good example. "I don't know if you like soccer," Vidirí said, "but in soccer terms, we say that Cabernet

Sauvignon, it hits on the post." The quality of Cab in Patagonia, in other words, is approaching the center of the proverbial target thanks to climate change.

Warming temperatures in Patagonia, Vidirí explained, means that Cab can be harvested earlier now than it was twenty or thirty years ago, before the autumn frosts sweep in. Back in the early 1990s, he told me, harvest for Cab in Rio Negro typically happened around the 25th or 26th of April each year—a risky time since frost generally became an issue around April 20th. This meant that Vidirí, Barzi, and their team had to carefully monitor the weather as April rolled on and make often painful decisions about when to pick. Too frequently, they were forced to get the grapes off the vine before they had achieved optimal ripeness to save them from an impending frost; given the decision between underripe and frost-destroyed grapes, the choice was clear if fraught. "Once the climate started to change," he added, "we have seen in the last years that Cabernet Sauvignon, Cabernet Franc, [and] Petit Verdot" have been better than ever: "Very fruity, very rounded." Because of warming temperatures over the course of the growing season, the grapes are developing more sugar than they ever have in the past, and harvest is starting earlier, too, which more or less avoids the threats posed by late-April frost. By the time the temperatures drop enough for frost to start speckling the landscape toward the end of April, the grapes have typically already been picked at optimal ripeness in most years now.

For all of the buzz and excitement around Patagonia today, however, it wasn't always perceived as being particularly promising from a winegrowing perspective. For a long time, Patagonia was thought of as simply not being warm enough even for the kind of grape varieties that thrive in cooler climates. It's located in the central and southern portions of Chile and Argentina, and while the popular reputation of the region conjures images of craggy glaciers and gargantuan snow-crusted pine trees, the truth is that Patagonia is so huge—it covers more than 400,000 square miles in total—and stretches over such a vast expanse of the continent (from around 38 degrees latitude all the way south to approximately 55 degrees) that it is a region of significant topographical and climatological range. So while its southern reaches are far too cold for the growth of much of anything, let alone wine grapes, its more northerly and even central portions are relatively temperate . . . and getting more so.

Back in 1998, *Wine Spectator* magazine ran a short yet prescient piece on the region that began, "At the southern end of the hemisphere, one of Argentina's most traditional wineries is pushing the frontier of grape growing. Bodega y Cavas de Weinert is developing a new project in El Hoyo Valley in Patagonia's Chubut province,"[3] close to 800 kilometers southeast of Rio Negro. "The first of its kind in the region, the Weinert project began as a 2.5-acre experimental plot of vines in 1990, and has now grown to a 360-acre estate. A total of 50 acres of Chardonnay, Merlot, Pinot Noir and Riesling are slated for planting this year, with plans to make top-quality varietal wines, according to Weinert's Swiss winemaker, Hubert Weber. . . . Both Weinert and Weber said they expect that the Patagonian climate will be better for Pinot Noir and Merlot than are the warmer areas, such as Lujan de Cujo, surrounding Mendoza—the heart of Argentina's winemaking."

How right they were. By 2004, just six years after that article ran, Patagonia had attracted the attention of some of the biggest names in wine. That was the year, for example, that Piero Incisa della Rocchetta, scion of the family that had founded Tenuta San Guida and its flagship wine, the iconic Super-Tuscan Sassicaia, launched Bodega Chacra, his vineyards and winery in Rio Negro, where Humberto Canale is located. Chacra's Pinot Noirs and Chardonnays have garnered plenty of acclaim, with their best bottlings earning scores in the nineties from top critics and commanding prices north of $100 per bottle.

Patagonia has, indeed, become an important region for the production of Pinot Noir, among so many other grape varieties. Today, in fact, as climate change is simultaneously making grape growing more challenging in some traditional parts of Chile and Argentina, it's making those very activities in Patagonia ever more promising. Still, there are challenges, and many of the issues facing regions north of Patagonia are starting to become problems there too; climate change, everywhere in the world of wine, eventually creeps in.

Along with the higher temperatures that Patagonian growers are benefiting from have come increasingly persistent winds. These can be beneficial and mitigate the potential threat of mold or fungus that humidity and rains often bring with them, but Vidirí told me that more and more often, the winds are becoming *too* assertive, which is not good for plant growth. He also noted the familiar problem of less rain but more intense bursts of it when it does fall, which poses issues of flooding, erosion, and more. Just a few weeks before we spoke in late-2021, he said, Rio Negro experienced a rainstorm that dumped forty millimeters of water in less than three hours. "It's too much," he lamented. "It's twenty percent of our average rains, and that's not good" in so short a period of time. "So, higher temperatures, higher winds, lower rains in general but high intensity *when* it rains, and long periods of drought. That is what is happening, and what will happen, in the next years in the opinion of the climate experts," he continued. "So we are rather worried about that, and working on that." He is also concerned that one of Patagonia's historical plusses, the massive rivers, which crisscross the landscape and have for so long provided a source of irrigation, could begin to dry up. Because of that, the amount of water diverted from them for watering the vines now has to be managed with an increasing level of precision . . . which given the unpredictability of the rains and droughts, is becoming more challenging and expensive.

The changes have been happening with a speed and an aggressiveness that few could have predicted a mere twenty years ago. Still, on balance, climate change has been beneficial to Patagonia. "I don't want to be a full [optimist]," Vidirí said, "because I think we are all worried about climate changes . . . and the retreat of the glaciers and things like that. But in terms of wines and grapes, I think we have been fortunate." Warmer temperatures in Patagonia, he added, "Have given us very fruity and very rounded and very agreeable wines that I think our customers are very [pleased with]. So again, I am worried about this. I think . . . everyone all over the world should be worried." He considered for a moment before adding: "But in Patagonia, in northern Patagonia, and [including producers and growers] who [are] going south, I think they will have very interesting experiences to achieve in the next years; I have no doubt of that." Among those who are looking farther south, they're heading deeper into Patagonia than previous generations could ever have imagined. The producer

Miguel Torres Chile, for example, now makes wine in Osorno, more than 900 kilometers south of Santiago. They also purchased, back in 2016, more than 1,800 acres of land in Coyhaique, more than 760 kilometers south of Osorno, in an effort to explore its potential for grape growing and winemaking.

That future may be coming faster than anyone anticipated a mere ten years ago. Temperatures in the Central Valley of Chile, the country's main winegrowing region, have risen with alarming speed. In Argentina, the story of rising temperatures is more or less the same. Grape growers and winemakers in both countries are increasingly looking south to Patagonia or up to the higher altitudes of their more traditional wine regions; as temperatures climb at lower elevations, planting at higher altitudes can help mitigate that.

I first met Sebastian Zuccardi, winemaking director for his family's eponymous brand, on a trip to Argentina in late 2019. Every time we sat down to taste and talk, it quickly turned into a master class in the ways in which terroir, microclimate, and viticulture all impact the final liquid in the glass. And while he of course realizes that the world of growing grapes is shifting—some of his high-altitude experimental vineyards are so far afield from where grapes have been grown in the past that they look like they're planted in a moonscape—he's nonetheless cautious when discussing climate change in Argentina, stressing the lack of "really strong information about the temperature. So it's difficult to say that the temperature is going up in a *scientific* way," he told me. Zuccardi is far more sanguine about climate change than most other producers I've spoken to, and it's important to understand where he is coming from: no progress on an issue as serious as climate change can be made without allowing for the full range of voices and opinions to be heard. It's also important to note that both Zuccardi and its sister winery, Santa Julia, are among the most forward-thinking producers I've ever visited in South America. They have robust social welfare programs for their employees, a serious focus on sustainability, and one of the most impressive soil-health programs I've seen. Sebastian Zuccardi's questions about the science behind how climate change is impacting the Uco Valley should not be taken as a doubting of the crucial importance of growing grapes and making wine in the most socially and environmentally responsible manner possible—something that every member of the family and company I spoke with is deeply passionate about.

"In our area, [the Uco Valley of the Mendoza region], apparently the warming is not so strong like in other areas," Zuccardi continued. He explained his perception of a more limited impact of climate change by citing a number of factors. "First," he told me, "because we are in the Southern Hemisphere, and in the Southern Hemisphere we have more water than land. In the Northern Hemisphere, you have more land than water. And this creates, maybe, [a situation] where temperature can be moderated by the oceans. So in the Northern Hemisphere it's clearly the warming situation, and here it's not so easy to see." That doesn't mean, however, that he isn't experiencing the issues being brought on by a changing climate. "One thing that we *are* seeing, and maybe could be part of this climate change, is that in the last eleven years, the average of snow that we have in the mountains is reducing. And as you know, we live in a desert, in a desert in high altitude, where the only way to cultivate is with irrigation. All the water that we receive comes from the snowmelt from the Andes Mountains. And in the last eleven years, the average of snow in the mountains has been smaller. And the glaciers have been reducing in size. Many people say that it's maybe a cycle only, and other people say that it's climate change. Really, there is no scientific information that you can say *this* is what is happening. But summarizing, in terms of warming, it's difficult to say that we are warming, but it's clear to see that we have less snow or less quantity of water in the mountains in the last eleven years."

Whatever Zuccardi or other grape growers and winemakers attribute that diminishing amount of snow in the mountains to, the subsequent lack of adequate water for irrigation has forced producers throughout Argentina to rethink how they work in the vineyard. The most important wine region in Argentina, for example, is Mendoza, a desert climate with a preponderance of high-altitude vineyards both heavily influenced by and reliant on water from the Andes. Interestingly, for all its growing renown around the world, Mendoza itself is not all that aggressively planted . . . because of water. "It's a tough situation," Zuccardi lamented, "because you know that we live in a place where irrigation is a must. You can't cultivate without irrigation. And Mendoza only cultivates 3 percent of the surface" as a result.

Visiting Mendoza for the first time, I was struck by the sheer scope of the place, by the incredibly varied terrain throughout the region. The wind-swept vistas of the Pampas are cut by the hulking silhouettes of the Andes at the edge of the horizon, their snow-capped upper flanks surprising in the heat of the Argentine spring. As you get closer to the mountains themselves, the geological forces that shaped this ancient land become more and more apparent, the folds and crenellations of the undulating terrain each hiding a new stratum of rock and soil. This is one of the aspects of Mendoza that makes it so appealing to grape growers and winemakers like Zuccardi yet also so frustrating; there is so much land here that would be perfect for growing vines but that just can't be planted to wine grapes in a financially reasonable manner because of the natural limits of irrigation. "When you have land, the owner of the water is the land. *You* are not the owner of the water—it's the land," Zuccardi said. In other words, no matter how promising a tract of land may be, if irrigation is too difficult or too pricey there, then its value is immaterial; without irrigation in Mendoza, there is virtually no farming of grapes. It's a naturally limiting factor, and an expensive one, at that. And with less and less water, increasing efficiency at all steps of the grape growing and winemaking process is the only real way to make the most of this precious and limited resource. "We can't change the quantity of water that we are going to receive, because we can't manage nature," he went on. "Nature gives us what it's going to give. But we *can* work on the efficiency of the conduction of the water."

To that end, drip irrigation has become the de facto method of bringing water to the vines, as it has in much of the wine world. (Prior to drip irrigation, flood irrigation was the standard.) The problem with that, perhaps inevitably, is a lack of money. Because while companies like Zuccardi's have plenty of it and the ability to invest in the kind of technologies—irrigation and otherwise—that are increasingly required for not just thriving but surviving in this new world, not every wine producer is in the same situation. Fortunately, the Argentine government's Ministry of Water has made serious efforts in irrigation infrastructure that will allow smaller wine producers to upgrade their systems without having to make massive investments. Still, Zuccardi told me, it's important to remember that Argentina is "a country [without] infinite resources." Add that to the fact that cities throughout the country are getting bigger, and the demand for water is exploding not just for agriculture but also for basic services in cities and towns. When I first visited Mendoza in 2010, for example, it was a charming, bustling city. During my most recent visit, in 2019, I couldn't believe how much it had grown not just in terms of the number of people going about their daily lives but also in the amount of new construction that had occurred over the course of the previous nine years. And as Zuccardi pointed out, limited water means that choices have to be made about where that mountain snowmelt goes. Providing water to the people—for washing, drinking, and more—will always be more important than diverting it for vineyards. Then again, there are calculations that have to be made in this regard too. The wine industry is an increasingly important part of the overall economy not just in terms of employment but also in sheer numbers. According to the *Wine Industry Advisor*, in 2019, "Argentina had wine exports valued at over $797.6 million representing 2.2 percent of the total global wine market."[4]

Fortunately, since the vast majority of water for irrigation comes from mountain snowmelt, it's easier and less expensive to irrigate the closer a vineyard is to the mountains; this is really a matter of infrastructure, and the shorter the distance that water has to travel from its source to its destination, the more affordable and easier it is to harness. In recent years, more and more vineyards have been planted and developed at higher elevations, which has been a significant help in this regard. Some of this is a response to the greater ease of accessing water there, but it's also thanks to the cooler temperatures that greater altitude provides. This is appealing from both a climate-change standpoint—chasing altitude or more extreme latitude is a common response to warming temperatures around the world—but it's also a result of the changing stylistic demands of the global market.

Until fairly recently, the most venerated wines in much of the world were the most powerful. There were exceptions, of course—the allure of the most sought-after Pinot Noirs from Burgundy have more or less always been predicated on their sense of elegance, their ability to transmit the ineffable truth of the place where they've been grown, as opposed to any overt sense of power—but in general, from the 1990s on, the wines that commanded the most money and generated the most buzz were wines of power. Much of this is thanks to the rise of the 100-point scoring system. In short order, that point system became the most widely used in the United States and throughout many parts of the world—England still largely relies primarily on a twenty-point scale, which while less granular, some would argue allows for a paradoxically greater sense of accuracy—and the wines that scored the highest began commanding the highest prices. This relationship between powerful wines and high scores became something of a feedback loop. In a blind tasting of dozens and dozens of wines, it only makes sense that the biggest and boldest would stand out; the fine-grained details of more subtle wines, as palate fatigue sets in, become harder to discern, while the more obvious aspects of a bigger wine, like the pleasantly weighty sensation of one with a bit more alcohol or the more overt fruit of a wine that has just slightly more residual sugar left after fermentation, come to the fore. For more than a quarter of a century, the world of wine headed in this direction.

Since then, however, the ability of a handful of critics to move entire markets like they once did has vastly diminished. Robert Parker sold a large chunk of the *Wine Advocate* in 2012, and in 2019, Michelin became the sole owner. Parker himself has stepped back significantly. *Wine Spectator* is still one of the most important American critical voices of record in the world of wine, and others, like Antonio Galloni's *Vinous* and James Suckling's reviews, as well as Wine Enthusiast and Wine & Spirits, have significant followings, too. But the rise of social media and aggregated, user-generated reviews of wine and the coming of age of a new generation of wine consumer that tends to value experience and connection more than the prestige of a high score have changed the landscape dramatically. The result of all of this is a world of wine in which the stylistic pendulum has swung—not entirely as there are of course exceptions all over the world, but certainly to a notable extent—in the opposite direction; today, many producers that once made their names on more overtly powerful wines are increasingly seeking out sites in which they can grow or purchase grapes that will ultimately, regardless of their sheer strength, produce wines of elegance, finesse, and detail too. And that means chasing cooler temperatures, poorer soils, and wider diurnal swings between daytime highs and nighttime lows, resulting in wines exactly like the ones that the higher-altitude areas of Mendoza are capable of producing, as opposed to valley-floor wines. And whatever Sebastian Zuccardi believes about the science behind the climate situation in Mendoza, his family's company, with their focus on mountain-grown wines—and, of course, sustainability—is brilliantly poised to take full advantage of current trends. In a place like Mendoza, with more and more producers looking to plant ever higher in the Andes, Zuccardi is well positioned to continue being the standout it has been for so long now.

For the Catena family, this need to plant higher has been a driving force for several decades. "Argentina's the perfect [place to] study climate change, because basically we addressed climate change, let's say, thirty years ago when we started moving all our vineyards to higher-altitude regions and further south," Laura Catena told me. She is a fourth-generation vintner, managing director of Bodega Catena Zapata—one of the most important producers in Argentina—and a part-time emergency physician in San Francisco (she splits her time between there and Mendoza). She also is the founder of the Catena Institute of Wine, an author, a Harvard and Stanford graduate, a polyglot, and one of the most public, prolific voices for both the wines of Argentina and the importance of respecting the environment. I'm pretty sure she doesn't sleep.

Historically, she went on, "Most of the vineyards in Argentina were in the eastern part of Mendoza, east of the city, which is around 2,700 feet elevation. . . . It's pretty warm; it's like Southern Rhône, maybe close to Languedoc," the sunny French region whose climate she compares it to. "And those are actually soils [in the traditional parts of Mendoza] that are a little deeper, retain a little more water, but it's warmer. And so my father basically figured out, if he wanted to make top wines that could compete with the First Growths and the best of Bordeaux for varieties [like Cabernet Sauvignon], and even, like, Chardonnay and Pinot Noir . . . he needed to go to cooler climates.

There were some vineyards in the [higher altitude] Uco Valley, but it was a much smaller region because there were frost problems, and also because it was further from the city. Oftentimes, places with cooler climate also have lower yields, and also the soils were better drained, so lower yields [again]. You don't always want lower yields unless you're specifically doing quality wines," she concluded, "and at that time in Argentina, people were just trying to feed the masses" with wines produced more with volume in mind than quality. As a consequence, Catena told me, "Malbec was being pulled out because it was considered . . . not high-yielding enough. And so it was being replaced by higher-yielding varieties" from which producers could make light, uncomplicated wines for the broadest domestic market possible. "Basically wine alcohol," she called it, "the beer version of wine."

But her father, Nicolas Catena Zapata, whom the University of California, Davis—essentially the Oxford of wine programs—has referred to as the Robert Mondavi of Argentina,[5] saw the potential in the country for higher-quality wines back in the early 1980s. "So basically my dad says, 'No, I want high-quality wines,' and he starts looking further south and higher altitude" in search of locations that will foster that. Catena Zapata began focusing on planting vineyards in the then-overlooked Uco Valley, which led, Catena told me, to great reviews from the key critics of the time. The message was clear: they had made the right decision. Other producers began to follow. "It used to be [that] eighty percent of vineyards in Argentina were in this warmer region" at the lower altitudes east of Mendoza. Catena said, "Now, eighty percent of the vineyards are in Uco Valley—anything new that was planted between the '90s and the year 2000. Basically, we moved to [a] cooler climate, we started doing it thirty years ago, and the rest of Argentina [followed] maybe fifteen years ago."

It was a smart move—Catena did as much as any producer to put Argentina on the world wine map—and in hindsight, incredibly fortunate. Today, as warmer temperatures and lack of rain, aside from the intense downpours I've already discussed in this chapter and others, have become increasingly prominent features of a climate change–impacted Argentine wine world, Catena has been working on better understanding the situation and searching for ways to mitigate it.

Water management has been key. "The study of how to manage water from a producer level, [and] also how to work with the government, is of constantly growing importance," she explained. Catena has made serious investments in this area as a result. "We have a person doing a PhD in water management at our winery," Catena continued. She added that at "the Catena Institute of Wine . . . we started by doing Malbec massal selections," or using particularly successful old vines as the basis for new plantings, which ostensibly allows a vineyard to be planted with vines that are better adapted to success in that particular terroir, that particular micro-climate. This, Catena clarified, allows for greater "genetic diversity, so we don't lose the genes that might help us in the future with climate change, like genes to fight viruses, genes to fight water stress, genes that can help vines do better with more intense sunlight or more heat."

Amazingly, learning how to succeed in a world increasingly impacted by climate change wasn't the family's driving force in the beginning. "We were looking for quality basically. We wanted to make age-worthy, Grand Cru–level wine." The climate change benefits came as a bonus. Today, Catena and the Institute are increasingly focusing on just that, and promoting its importance to other producers, as well. "My dad, he has a PhD in economics. I'm a medical doctor. I come from research. We like to do research; it made sense," she said. "Now, I think that basically, the future of wine depends on every winery doing some research. Every winery needs to understand what's going on in the scientific world" as it pertains to climate change.

Still, understanding is only the first step. Real change is always difficult without the money to implement it. Which is why Catena is taking matters into their own hands.

Wine-producing countries around the world are dealing with an array of climate disruptions. But in Argentina, Catena told me, "We've got all of them. We had the warmer climate and [so] we moved to cooler climate. We have the water issue . . . [so] we moved from flood irrigation to drip irrigation in the last thirty years. So we've already seen how much water you can save by not doing flood irrigation. *Nobody* does flood irrigation. Well, some places do, but very little anymore. We've also gone through [the question of], how do you transform a whole region from one practice to another, when drip irrigation's very expensive? Most people don't have [enough] money from one whole harvest to do that. And what ended up happening is that . . . we'd talk to a grower and we would fund their drip irrigation. And then we discount it, we basically give them credit, so there's financial ways that you can do this. And," she went on, "the other reason why we're ground zero [for looking at the impact of climate change on wine producers around the world] is that our government has very little money. I think that there are places like Australia, where they have this big tax and they actually really have money. They have money to promote their wines, they have money for research, they have the Australian Wine Research Institute—that's amazing. So I think in Argentina, it's like a really good model for how you figure things out, having [less] money from the government, or very little," aside from the efforts of the government's water ministry that Zuccardi mentioned. She paused, then added: "So what I'm saying is, Argentina has had all the problems, has very little resources, and I feel like we're going to figure it out because . . . there's also freedom to do things," to come together and find a solution. She and her family's company are leading the way, putting action and money behind their words. And the impact has been significant, both for them and for the growers they rely on who wouldn't necessarily be able to afford climate change–mitigating efforts on their own.

In the world of wine, there is a lingering perception that it's the smaller, scrappier producers who have the greatest focus on sustainability, environmentally respectful production in the vineyard and winery, and overall sense of corporate responsibility. And while it may have been true that historically that it *was* the smaller players who could afford to train their focus on what were usually perceived as less bottom line–determining initiatives like sustainability and organics, that entire calculus has not just changed in recent years—it's been completely flipped upside down. Today, in fact, it's often the biggest producers who are leading the way in the sphere of sustainability, and because of their size, they're often having the most significant impact. Catena and Zuccardi are good examples in Argentina; VSPT Wine Group is an equally important one in Chile.

A decade ago, Barbara Wolff Göpfert's position as chief of corporate affairs and innovation for Viña San Pedro—Tarapacá Wine Group was just as much about perception as anything else—a nod in the direction of responsible practices whose real-world impact was secondary to the message it sent. The company is one of the twenty largest wine producers in the world, and among the twenty brands under their umbrella are both the kind of wines that can easily be found at supermarkets as well as smaller-production bottlings that have earned the respect of critics and collectors alike. Yet what is now a guiding principle of the company's work—leveraging innovation and technology in the fight against climate change—started out as most revolutions do, with a handful of passionate people who didn't want to accept the status quo, with a younger generation that was less willing to accept the received wisdom of the past and enthusiastic about challenging the ways things had always been done versus how they believed they actually *should* be done.

"I guess at the beginning, it was just sort of a couple of employees within VSPT having this impulse and this spirit and this commitment" to help the environment, Wolff explained. "But after a few years, it started to become a sort of value for the organization. . . . I mean, having a commitment, having a policy, having a strategy, having a road map on how to become more sustainable, [it all] became an added value, not only for the [people in the] organization, but also for the business." In other words, what started off as a passion project for a few committed employees has turned into one of the most significant, forward-thinking efforts in sustainability in the entire South American wine world. And it's Wolff's job to manage it all.

After a few years of work from the small group, the higher-ups in the company started to take notice. "The first level of executives started to realize that this was actually adding value to the business as well," she went on. "Maybe ten years ago, it was a way to differentiate yourself—you know, having a higher standard, or maybe a different approach on the business, and the way you do business, because this is not only about the environment, it's a holistic approach, right?" As word about the efforts of this small group reached the upper echelons of the company, it was initially seen as a good marketing opportunity as well as a way to keep their younger employees happy and engaged.

"But then it became a business tool," she went on, "and we started to understand, *everyone* started to understand, that this was the way to go." Given the scale of the company's production, executives soon realized that investing in measures to mitigate their impact on the environment, especially as the effects of climate change became more apparent and destructive, would be beneficial in the long term—for their land, for their wines, and for their bottom line. Which is why Wolff's position was created.

Today, the results of their efforts have completely changed the business practices and corporate culture of VSPT Wine Group . . . and because of the company's size, of much of the Chilean wine firmament. "It's huge how much value it has added to the organization, and how much prouder people feel within the organization when you talk about excellence in your operation—when it's not only about the wines you produce, but it's also the *way* you produce them."

This is just as much a consequence of generational differences as anything else. For VSPT's younger employees, Wolff told me, "It's not only about where do you work or how much do you gain or how far are you from home. . . . Today it's, 'What's the purpose behind your company? What's their sort of reputation?'" Among this younger generation, she pointed out, "No one wants to work somewhere that actually is harming the environment, or is not doing things right." She paused before adding, "My parents' generation, if you got a good job working for the cigarette company, you worked for 35 years, you took your pension, and that was it, whether or not you agreed with them. . . . And that's a real paradigm shift now."

That kind of massive change in the perception of what work is actually for is a result of both generational differences as well as the increasingly dramatic impacts of climate change; it's no longer a hypothetical, and year after year, the most dire predictions of a decade or so ago become more and more viscerally felt . . . and often exceeded. And it's the largest companies, with all of their land, facilities, and capital that can tend to have the greatest positive impact. "We have to understand that global warming doesn't necessarily only need to be translated to having higher average temperatures," Wolff explained. "It also means having more severe droughts or long-term droughts. It has to do also with more severe climate *accidents*." In Argentina, where several of the VSPT brands are based, hail has been a major issue, as has rain—both too much of the latter and, just as often, the lack of enough. "I can't recall whether it was this year or last summer," Wolff said when we spoke over Zoom during one of the COVID spikes, which made international travel close to impossible in late 2021, "but we had some, I don't know what, 300 millimeters of rain in 24 or 48 hours. So it's huge, it's unbelievable, it's something you can't even picture in your head," she said. "When it comes to Chile, one of the threats we are facing right now—[and] it's very profound—is long-term drought within the central and southern area. So water scarcity, it's an issue today. . . . We have to include this as a long-term threat that we're going to have to manage in a different way."

For a company the size of VSPT Wine Group, managing the threat of both too much rain and not enough, on both sides of the Andes, is a matter of tens of millions of dollars in gains or losses over the next several years. So, too, are the ways in which they diminish their carbon footprint, increase the sustainability of their vineyards, and become overall better actors in the wine world. Fortunately, they have the resources to approach the problems in a very different way. And their decisions today will likely have an impact on the South American wine world for generations to come.

Their efforts have been substantial. "When it comes to . . . action for climate change, one of our main goals here, or commitments, is to become carbon neutral at the latest by 2050, which is aligned to the UN's Race to Zero commitment," she told me. "In order to do so, we joined IWCA, the International Wineries for Climate Action, in late 2019. And since then, we have been working together in establishing our baseline, [our] carbon-footprint baseline . . . and then committed to continuous reduction, having the next milestone in 2025 with a 25 percent reduction, and 2030 with a 50 percent reduction. And the beautiful thing behind this is that we have a road map, and we have a sort of calculator where [we can ask of each step along the way], 'What if we do this, and how does it affect or how does it reduce?'" Wolff is also overseeing VSPT's efforts to work with glass manufacturers to lower the weight of the bottles they use, with the suppliers of the chemicals they employ at various steps of wine production, with the shipping and logistics partners to make sure that their wines are transported in as environmentally conscientious a manner as possible. And she added, "We established ourselves to become 100 percent renewable in our electrical demand by the end of this year [2021]. I think that we're going to achieve 97 percent. The 3 percent rest is going to happen in the first quarter of next year.

"But," she clarified, "it's not about buying renewable energy or just making sure the energy is renewable. We have committed ourselves to also generate and consume our proper energy through three different sources. We biodigest the whole organic waste from our harvest—so we have two big biodigesters next to one of our facilities, which was very innovative, and we launched that in early 2016" after a $5 million investment. "There was no winery that was using its waste to produce energy. The beauty about this is it's not only energy that we produce. We also use the outcome after this, and we put it together to the vineyard as a biofertilizer. You can't imagine how beautiful that looks." Wolff is also leading the initiative to "work together with our main suppliers so they also improve their emission factors" in the production and transportation of what they produce. Finally, she said, "We're going to start measuring the level of healthiness of our soils. . . . We're going to measure the level of organic material in different spots in different vineyards we have throughout Chile. We will start with this pilot in Chile, with [a] trial in Chile. And then, we will also include into this metric the potential of carbon sequestration by adding . . . more organic material." Given the scale of land that VSPT Wine Group works with and the sheer volume of wine they produce, the massive amount of bottles, labels, corks, chemicals, electricity, and resources throughout the grape-growing and winemaking process each year, as well as the transportation of it all to locations around the world, the short- and long-term effects of their efforts promise to be huge. It's a textbook example of how a large company can have the greatest impact when it does the right thing.

Throughout Chile and Argentina, climate change is being felt in both subtle and more extreme ways, and producers across the spectrum of size and influence are being forced to pivot and modify the ways in which they respond to it. Some are moving south, deep into Patagonia. Others are heading higher in the mountains, seeking out altitude. Larger organizations like Zuccardi, Catena, and VSPT Wine

Group are leveraging their size and resources to help make climate change–mitigation efforts easier (and in some cases, simply possible) for their smaller counterparts. Virtually everyone, however, seems to be responding, whether out of altruism or necessity . . . though the line between them, given the situation these days, is growing increasingly blurry.

The best way forward—perhaps the only way—is together. "It's a very associative and collaborative way of finding a solution where everyone wins," Wolff summed up. "And that's actually the way we need to tackle climate change. It's not about putting the whole stones into your bag and making it heavier; it's about finding solutions in the most collaborative way. . . . We have to stop thinking about looking into our [own] operation, [and thinking] it's just about us. That's not the way it's going to work anymore." It's about everyone—the problems, yes, and also the solutions.

7

DEEP FREEZE IN HILL COUNTRY

The team at Spicewood Vineyards should have finished harvesting their estate vines by the time I arrived in mid-October. The juice should have been burbling away in its stainless-steel tanks or oak barrels, the yeast feasting on the sugars that had accumulated over the course of the growing season, the team punching down the cap of skins that rose to the top of each one in an effort to extract as much flavor and color and tannin as possible. Afterward, in a more typical year, the young wine would have been racked from one vessel to another to separate it from the solids that would have accumulated over the course of the winemaking process. From there, it would have aged for six or twelve or eighteen months before being bottled, sold, and savored by Spicewood's growing base of fans throughout Texas and across the United States.

But not this year. Though the team at Spicewood will be releasing wine from the 2021 vintage, all of it will be produced from grapes sourced from their vineyard partners. In a typical year, they grow Tempranillo, Syrah, Grenache, and Sémillon on their twenty-eight acres of estate vineyards that stretch out from the no-frills but very comfortable tasting room toward the wide horizon off in the distance. But Winter Storm Uri, whose full fury began to be felt on

February 13th of that year and didn't officially let up until the 17th (though other systems continued to impact the state), rendered Spicewood's vines incapable of producing enough fruit or the quality of fruit that its wines require. Luckily, Spicewood was able to source grapes from a number of its Hill Country vineyard contacts, some of which fared remarkably well. The vines that form the basis of their wines in the Texas High Plains, more than 200 miles away, generally produced brilliant fruit. But at Spicewood itself, "It was pretty devastating," owner Ron Yates told me. Sure, some of the vines produced a couple of berries, but not enough to justify the time, effort, and expense to pick it—not to mention the impact on the vines themselves after the trauma of the deep freeze. "We could've got in there and had some secondary [or] tertiary fruit . . . but that was an ultimate decision of ours, to just drop the little bit that was there, and instead of trying to make minimal amounts, let the vines get healthy for next year."

It was the grape-farming equivalent of amputating a mangled limb: not at all a decision that you ever want to have to make but ultimately the one that in the long term, and despite the immediate pain, is the necessary one. Yates realized that given the unprecedented set of circumstances, it was the only rational choice.

Grape growing and winemaking in Texas have always been challenging propositions. Unlike the most famous wine regions in the United States—Napa and Sonoma are good examples—Texas doesn't benefit from the kind of generally even and predictable weather patterns that have allowed its California counterparts to achieve greatness. When I asked Yates what a typical growing season was like, he scratched his thick beard, ran his right hand through his Allman Brothers hair, and laughed. "Typical's a really good word," he said, smiling slightly and sinking his hands into the pocket running horizontally across the front of his hoodie.

"Typical" in Texas, it turns out, is an awfully slippery subject. The state may be perceived by outsiders as more or less unrelentingly hot in the summer, cool yet temperate in the winter, and aside from a few Texas-sized thunderstorms in the spring and autumn, more or less a place of predictable weather patterns. But that's just not the case, and everyone I spoke with there who's involved in the wine business pointed out that historically, Texas, across its wide swath of land, is buffeted unpredictably by factors that are often difficult to account for—the intersection of storms in the Gulf of Mexico, the jet stream dipping down from the lower Midwest, mammoth rain and hail events developing across the broad, flat flank and riding through the western and central parts of the state, all of which have always tended to make grape-growing a tricky business in Texas. This is why the most successful vineyards and wineries have been the ones that are the most flexible, the most willing to pivot and try new things, even if they go against the received wisdom of winemaking and the market.

Climate change, however, is making the old sense of unpredictability seem logical in comparison.

Winter Storm Uri was way beyond the pale—literally without precedent. According to Houston Public Media, 210 people lost their lives as a result of the storm[1], and based on numbers compiled by the University of Houston, "More than two out of three (69%) Texans lost electrical power at some point . . . for an average of 42 hours, during which they were without power on average for one single consecutive bloc of 31 hours, rather than for short rotating periods."[2] In addition, "Almost half (49%) of Texans lost access to running water during this week period, with the average Texan who lost running water without it for 52 hours. During this same time frame, the average Texan with running water could not drink it for an average of 40 hours." With that kind of death and destruction, the loss of a winery's ability to grow and use its own grapes may seem like a relatively small matter, but the economic consequences are substantial . . . and potentially long ranging.

The weather system itself, which originated in the Pacific Northwest and arced its way across the United States, unfolded like a slow-motion train wreck. According to the National Weather Service, "It all began Wednesday, February 10th when a cold front moved through the area bringing the first surge of cold air into the region. With this cold air in place, lingering precipitation the following day fell as sleet and freezing rain across the northwestern counties."[3] The next morning, on February 11th, the first of what would become a cascade of winter weather advisories were issued, largely the result of ice on the roads. At the same time, an Arctic cold front was on the march, and there were already concerns that it would find its way to Texas by the end of the following weekend. It was another case of Texas being impacted by seemingly discrete weather systems that, unfortunately, came together over the state in exactly the wrong way. The weather postmortem continued: "On Friday, February 12th, a Winter Storm Watch was issued for the entire region for Sunday in anticipation for the potential snow, sleet, and freezing rain that this Arctic front would bring. A Winter Storm Warning ended up getting issued on Saturday, February 13th for Colorado, Austin, Waller, Montgomery, San Jacinto, Polk counties and for counties north as sleet and freezing rain formed ahead of the approaching cold front. The counties that remained in the Winter Storm Watch . . . got upgraded to a Warning for Sunday." As that first system was dumping its wintry precipitation on Texas throughout the weekend, the dreaded Arctic front passed through on Valentine's Day. The arrival of that front, according to the report, "served as the turning point from a significant winter storm [that] preceded the front to the *historic* winter event that would eventually unfold."

By the time Valentine's Day had arrived, "Every square inch of Texas was in a Winter Storm Warning. Snow, sleet, and freezing rain began to encroach into Southeast Texas Sunday afternoon, and then increased in coverage and intensity overnight Sunday night into Monday." The scene was turning biblical: Thunder snow and thunder sleet rumbled down from the sky, and the precipitation and plunging temperatures became so extreme that "roads began to become impassable through the region Sunday evening due to ice and snow and some would not become safe until Friday," the report went on. "Temperatures crashed down Sunday night behind the cold front with much of the area getting down into the teens or single digits with wind chills down into the single digits or even below zero. Because of these conditions, a Hard Freeze Warning and a Wind Chill Warning (the first in our office's history) was in effect Sunday night / Monday morning. A combination of the icy conditions and extreme cold temperatures caused widespread power outages that would last for the next several days. The wintry precipitation continued through Monday morning with storm total snow / sleet accumulations being around trace along the coast, around an inch near the Houston Metro, and up to three to six inches up across the north. The extreme cold temperatures not just persisted through Tuesday morning, but dipped down even colder and produced the coldest morning of the event: the City of Houston went down to 13°F, Galveston down to 20°F, and College Station bottomed out at just 5°F."

There was still more to come: Another system was heading toward the battered state, more storm warnings were issued, and temperatures remained frigid. "It wasn't until 9am Saturday morning," the National Weather Service concluded, "that the last Hard Freeze Warning would expire for this event."

By the time the storm was over, it had lasted for nearly nine days. In its entire recorded history, Texas had never experienced anything like this, and the damage the storm had left in its wake approached incomprehensible proportions. The cost of its immediate impacts and the issues that followed close in its wake climbed into the billions of dollars. And yet, according to the University of Houston, while "more than two-thirds (69%) of Texans agree that due to climate change Texas is more likely to be adversely affected by severe weather than 30 years ago . . . [only] 61% of Independents and 46% of Republicans feel the same way. (Among Democrats, that number is 95%.)"[4]

As the climate becomes increasingly unpredictable, causing storms of a power and magnitude previously unseen, the damage will continue to grow, no matter how vociferously its deniers stick to their positions. And for people like Ron Yates—as well as his fellow winemakers and grape growers—who don't just find themselves more frequently at the mercy of the elements but whose entire livelihoods are based on somehow managing them, the repercussions will continue to grow in size, scope, and destructiveness.

Ron Yates is a charmer. He's the kind of guy who can spin the proverbial yarn better than some octogenarian nineteenth-century grandmother sitting at her spool. The joy in engaging him in conversation is trying to predict the directions in which his stories will pivot and turn, the little sidebars that may seem unrelated initially but that usually, somehow, form the crux of what he's saying: He's smart as hell, but wears it lightly. His *sui generis*, almost hippy-inspired aesthetic—he may come from a long line of more taciturn, traditional Texans (his grandfather, Tommy Joe Yates, was a strait-laced, "hardscrabble" man, as Yates described him to me, who rationed out his words more carefully than a small-town banker does loans)—makes Yates himself look like he could have been teleported to Texas Hill Country from Laguna Beach circa 1978, the year he was born. The overall effect is equal parts disarming and winning.

The way Yates tells the story, he convinced his family to buy Spicewood Vineyards the old-fashioned way: he sat down with his parents and sister and over a great deal of wine, they made the momentous decision. Spicewood had been around since 1992, and while the wines coming off of its seventeen planted acres of vineyard were plenty good, they weren't nearly approaching the potential that Yates knew they were capable of. That's what sold his family—the fact that he saw greatness hiding in the soil. The only issue was the classic one: money. While they had enough to fund the purchase of the winery and the vineyard, there were also two adjoining parcels of land that the owners wanted to sell, which brought the price over their budget. This is where his grandfather—who Yates used to call "Good Guy" as a child (the story goes that he couldn't quite pronounce "grandpa" as a baby; it came out sounding like "good guy," and the moniker stuck)—came in.

Tommy Joe was a quiet man, cut from the classic cloth of Texas Hill Country: a fifth-generation Texan who could be relied on, a man whose unflagging loyalty to his family and friends was never in doubt but who was so deeply rooted in his time and place that Ron's differences remained perplexing to him throughout his life. It never adversely affected their relationship, though, and they remained close until his death in 2010.

The morning I visited the Spicewood tasting room, Yates showed up in shorts and a hoodie, his shoulder-length brown hair parted just off the middle, his beard's bushiness attenuated by its well-kept symmetricalness. His playlist was an impeccably chosen amalgam of blues-heavy country, and he was able to speak just as passionately of his roots in this mythical part of Texas as he was his days running around as a college kid chasing the local girls during his time studying abroad in Spain, days he spent falling in love with the local ladies and the great wines of the country. The son of the family he was living with grew grapes in the venerated Spanish region of Ribera del Duero, and the nuts and bolts of that kind of farming, as well as the lifestyle around it, spoke to him. Even when he came home and after stints in law school and in the music business (he cofounded High Wire Music), he just couldn't shake the wine bug that had bitten him. Nor could he have been expected to: the landscape of Texas Hill Country kept on reminding him of Ribera del Duero. Perhaps the universe was telling him something.

In fact, that was one of the keys to his convincing family to go in with him on Spicewood: the dry, dusty land cut through with rivers; the hot sunny days and brisk nights; the alluvial soils speckled through with clay and limestone—all of it reminded him of Ribera del Duero, where some of the best Tempranillos in Spain are grown. There was just one problem: between them, they didn't have enough money to finance the purchase of the entire property, including the adjoining parcels, despite the fact that Yates had convinced the original owners, Madeleine and Edward Manigold, that he would be a good steward of the land and all that they'd toiled to accomplish. Yates had worked with them that summer and into September for harvest, and the mutual respect that blossomed seemed to seal the deal.

Eventually, Tommy Joe told Ron that he'd buy it for him. Actually, that's not quite accurate—he'd buy it, keep the ownership of it in his own name, and lease it to Ron for ninety-nine years at a grand-total cost of $1. That way, if Yates started making what he told me his grandfather called "foolhardy" decisions, he could take it back. It was more of a message than anything, seeing as he was getting on in years at the time, but his point was clear: Ron best not mess this opportunity up. Plus, Yates told me, Tommy Joe saw this as the moment for his "long-haired, hippy grandson to get out of the music business and into something more respectable, like agriculture."

It was a sound investment. Since 2007, Spicewood has become one of the most lauded producers in Texas Hill Country and the High Plains, producing what are widely considered to be some of the finest Tempranillo-based wines (and others too) in a state increasingly full of them. They've had their fair share of challenges, of course—the perceived glamour of the wine business is inversely proportional to the level of difficulties it poses to even its most accomplished and experienced practitioners—but even the major replanting of ten acres because of a Pierce's disease infestation (this is a relatively common problem, caused by the bacterium *Xylella fastidiosa*, spread by the glassy-winged sharpshooter, and impinges on a vine's ability to conduct water) didn't stop their progress. They just replanted, took mitigating measures once they did, and moved on. Since then, the family has also started Ron Yates winery in the town of Hye, Texas, and their Friesen Vineyard 2017 Tempranillo, from the Texas High Plains, won the prestigious award for best red wine in Texas at the 2021 Houston Livestock Show and Rodeo Wine Competition, one of the largest in the country and by far the most prestigious in the state.

In a remarkably short period of time, Yates has managed to achieve what he convinced his family he would do: produce wine that would rival the quality of the ones that made him fall in love with it in the first place, back when he was a college kid abroad in Spain. Yet he's never forgotten where he came from and the people who have made it possible: One of his top reds, a Tempranillo-dominant blend, is called "The Good Guy." It's named after Tommy Joe, and the back label prominently features a photo of him, hair parted off to the side and swooping gently upward, Kennedy-esque Ray-Bans blocking out the Texas sun, and a smile on his lips. Yates told me that it's one of the few photos of his grandfather in which he's not wearing a cowboy hat. That back-label picture is an homage not just to him but to the generations that came before.

The last time Yates saw his grandfather was 2010. He and his cousin had just come in from working on the land that his grandfather had helped make possible, covered in dirt and exhausted from a long day of labor in the vineyard. It was proof, for Tommy Joe, that he had made the right decision in trusting his grandson's instincts and enthusiasm: Here he was, a soil-caked, respectable man of the land, just as he wished for him. The first vintage of wine that Ron produced was from the 2011 harvest, a year after Tommy Joe had passed. He never had the chance to taste the result of all that hard work, but he saw with his own eyes what he had helped to create.

Vintage after vintage, Yates's wines improved, gaining greater and more acclaimed traction with each passing harvest. Everything seemed to be going exactly as he'd hoped it would. And then, in the second week of February in 2021, everything Yates had built was under a threat unlike anything he or any of the generations of the Yates family before him could have imagined.

The Thursday before Valentine's Day, Yates began obsessively checking his weather apps, which is what he always does when a storm is forecast to roll in. "Being a farmer, I have about fifteen different weather apps on my phone," he told me. He wasn't overly concerned. "Everything was saying, 'Oh, it's going to be cold. It's going to snow a little bit, but it's not going to be that bad of a deal.' But I have this one weather app called Dark Sky, which was saying, 'Hey, it's going to be below freezing for like five days,' and nothing else was saying it." In the past, however, that particular app had been more accurate than others when it came to predicting Hill Country weather, "So we started kind of paying attention to it. And the closer it got, you could just see the perfect storm of the high pressure moving . . . and the cold air kept dipping farther and farther and farther. Man, I tell you, we can deal with cold, but we are not used to cold for that long, [for] extended periods of time." And it's not just the people who call Texas home who aren't used to that sort of weather; the entire infrastructure isn't built for it. "I think the city of Austin has one or two salt trucks to salt the roads," he added. "We just don't deal with that stuff. And it was very out of the ordinary for a lifelong Texan."

Still, he kept on hoping for the best. As a farmer, Yates's tendency is to obsessively follow the forecasts and to take them seriously but also to look at what's happened in the past to get a sense of context, a framework for understanding what was likely to happen. Unfortunately, in this era of extreme weather caused by climate change, the past is no longer—or at least not as often—an accurate predictor of the future: everything is just *different* now.

"Normally," he went on, "if there's going to be an event like that, we'll get together [and] figure out what we're going to do." Sometimes, that means turning on the tall wind fans that are an increasingly common sight on vineyards. They leverage the fact that colder air sinks, whereas warmer air rises, and when they're turned on, they blow the slightly warmer air onto the vines, which even if it's just a degree or two higher, can be an effective preventive measure against frost. Some vineyard managers in Hill Country will set out hay bales at the ends of their rows of vines, set them on fire, and allow the heat to protect the fruit. These two techniques have been reliably protective, especially considering that freezing temperatures in this part of Texas generally didn't last all that long and rarely involved the mercury dipping below the high twenties. But the day before this particularly unprecedented storm descended, Yates's trusty weather app kept on getting more and more ominous. "Well, it's going to be sixteen. Oh, no, never mind; it's going to be ten. Oh, the third or fourth night's going to be minus three. I mean, honestly, we had never seen anything like it." Every time he checked, the forecast seemed to become more dire. "I was probably not the easiest person to be around in my house for a couple days," he admitted. "And we probably consumed a few more bottles of wine that weekend than I probably should have. . . . But it was just kind of a cross your fingers and hope it all works out. At that point, there was really nothing that we could do."

So he waited, checking his phone, pacing, popping the cork from another bottle of wine, checking his phone again, and putting it all on repeat. It was a feedback loop of fear and wine, fear and wine. At one point, he considered hopping into his four-wheel-drive vehicle, which can plow through most weather, but at the urging of his wife Jessica, he didn't head out from their home in Austin to the winery. It's an hour's drive in good weather, but given the severity of the storm and the lack of plows and salt—this is central Texas, after all, not exactly part of the snowbelt—it would have taken him three or four times that long . . . if he could make it at all. So he called the eighty-something couple that lives next door to the winery for updates and found out that though they had lost power, it had only been out for six hours or so, which meant that the damage to the wines still resting in the winery itself would likely be minimal. (He was right. Because of the insulation of the barrel room, its underground location, and the tendency of liquids to hold their temperature more assiduously than air, the temperature stayed at a perfect fifty to fifty-five degrees throughout the event.)

But he didn't know that for sure at the time, and by Saturday, Yates couldn't wait any longer. What did the vines look like? Had any pipes burst? When he finally made it to the winery, he was prepared for the worst but found that for the most part, the physical damage to the building and its plumbing wasn't too bad: A couple of burst pipes had caused water to spew from the walls, but the drains in the floors worked perfectly. A few ceiling panels had to come out in order to access the broken pipes and prevent mold from forming, but aside from that, the damage to the physical structure was minimal. "Initially," he told me, "I walked out of the weekend really relieved that we didn't have a whole lot of damage to the facilities. But it was just at that point I was like, 'What happens with the grapes?'"

By mid-February in most years, the vines at Spicewood are a week or two away from budding. There's usually a "warming event," as Yates called it, the first or second week of the month, and that kick-starts the vines; the first tentative hints of green life can start to be seen. After especially cold winters, the vines don't start pushing buds until around the first of March. But after this storm and the deep freeze that lasted for so many days, there was no sign of life at all on Spicewood's estate vines. The Ides of March passed, then the first day of spring. World Meteorology Day, on March 23rd, ironically came and went. Still nothing. "There was about two weeks there where we were like, 'Well, everything might be dead.' And then finally, you'd see Tempranillo push a little leaf bud, and you'd see it over here, and you'd see the Merlot . . . and finally, by about April, everything was out. But it was way later, and everything was much slower to really go." The growing season would be shot. Spicewood's vines didn't develop a full canopy of leaves until July, which impeded their ability to photosynthesize sugar from sunlight. The lack of canopy also meant that the vines and whatever grapes they did produce weren't shielded from the sun's rays. On top of that, the delay in the development of an adequate leaf canopy also prevented appropriate fruit growth and development: no sugar means no energy to push fruit. "I mean, it spirals," Yates said, shaking his head. Once those dominoes start falling, there's really no way to stop them.

When I visited in October 2021, harvest was wrapping up all over Hill Country. But at Spicewood, as at some other vineyards nearby, he picked no fruit from his estate vineyard. Not a single grape would be harvested for wine.

Climate change is often referred to as global warming, but that really doesn't adequately explain what's actually happening. In fact, the more wine and spirit professionals I've spoken with, the more I've heard pushback to that term. "I actually prefer to think of it as 'climate weirding,'" Julie Kuhlken, cofounder with her brother, David, of Pedernales Cellars in Texas Hill Country, told me over cocktails at Camp Lucy, the luxury boutique hotel in Dripping Springs, Texas. I was in town with a number of colleagues to taste the wines of the region and to speak with producers and grape growers about how climate change was impacting their day-to-day work and ongoing projections for the future. We were having drinks with a handful of producers and journalists in a space that looked like it had been beamed in from the other side of the planet—which in essence, it had been. The wood and decor for the bar area had been brought over, piece by piece, from Vietnam, after having been labeled there and then disassembled, packed into shipping containers, sent over the ocean, trucked across Texas, and finally reassembled right where we were sitting, more than 8,000 miles away. Looking closely, I was able to discern the markings on the beams and joints indicating how it was put back together, and the effect of it all caused a bit of cognitive dissonance. The excellent cocktails made that both better and worse, as they are wont to do. Kuhlken told me that before she went into the family business—her parents purchased the land that would become Pedernales back in the 1990s—she spent her time in academia. Even after she got involved with the winery, she kept her toes in the philosophical waters, working as an adjunct professor at Concordia University Texas and as an assistant professor of philosophy at Misericordia University. As recently as 2013 and 2014, she served a term as the president of the Southwestern Philosophical Society. Her work in that field has continued up to today, despite the pressures of handling the marketing and hospitality for Pedernales. She was recently hired to write a chapter in the next installment of the Blackwell Philosophy and Pop Culture Series; her work will consider the aesthetics and marketing of the rock group Queen through the lens of German philosophy, with a focus on Adorno and Nietzsche. She has a preternatural ability to see the often confusing and seemingly conflicting connections among disparate strands of a topic and isn't afraid to share her opinions. Mapping a conversation with her is like the linguistic equivalent of one of those renderings of a neural network: it splits off in a million directions but always comes back to the central theme. Much like conversing with Yates, it's invigorating. "I mean, it seems to me that 'climate weirding' is a more accurate term because that's what's happening. Things are getting *weird*," she said, sipping from her glass.

This is an opinion that I heard a thousand times, a corollary to the one that blames the term "global warming" for so much of the political pushback to it. Earlier that day, at the tasting room of Duchman Family Winery, winemaker Dave Reilly told me that soon after he was able to receive guests again after the February storm, some of his visitors started joking about the massive snow, ice, and freezing event, saying things like, *Well, how do you like that global warming now?* or *Bet you'd have loved some of that global warming during that storm, right?* Reilly, of course, sees the flaw in that sort of logic——climate change is most likely *exactly* at the root of the storm—but the fact that the term "global warming" has been applied to a climatic phenomenon whose impacts are more about unpredictability and extremes than simply increased heat has given doubters and deniers all the ammunition they need to make false claims about its veracity.

But on the ground, in the vineyards, the impacts are impossible to ignore, much less deny—even in such a traditionally conservative place like Texas. (Which, of course, is politically shifting in parts.) The Kuhlkens have been growing grapes on their land for twenty-six years, and until recently, they hadn't experienced a single hailstorm that impacted their work. Now, Kuhlken told me, "We've had three in the last seven or eight years. That means we have to change what we're doing. In fact, we *are* changing. We've just bought hail netting. We had never bothered with hail netting, and now we're like, 'No, that makes no sense.'" It's a significant expense when you consider that Pedernales has over seventeen acres of estate vineyard land under vine in the Bell Mountain AVA, an American Viticultural Area within the larger Texas Hill Country. But the other option is to not net the vineyard at all and risk potentially vintage-ending damage. And that's just hail. There's not much that can be done for extremes like the storm of February 2021 except hope that your vines somehow make it through.

Fortunately, grape-growing in Texas is perhaps likely to be more resilient than it is in some other places simply by virtue of the historically inherent unpredictability of the climate and weather there. In fact, the entire character of the industry has been built on a foundation that accepts the fact that it's not Napa Valley or Burgundy, Kuhlken told me, places where a single grape variety—or handful of varieties—is discovered to be uniquely well suited to that land and climate and the minute differences from one block of vineyard to another, one flank of a hillside to another, and can be explored and expressed in the wines. "We'll never be doing that," she said. "Almost everything is about blending because every year is going to be different."

Those differences, however, historically occurred within certain accepted and relatively reliable parameters. Climate change has blown up that continuum, which means that even knowing which grape varieties to plant has been thrown into question. For much of modern winemaking history, the most lauded regions and appellations have been associated with specific grapes, thanks to the relatively predictable character of the climate in those places, and of course to other aspects of their various terroirs. For centuries, Pinot Noir and Chardonnay have dominated vineyard plantings in Burgundy—records going back to the days when literacy was very much the purview of the aristocracy and the Church show Cistercian and Benedictine monks following how one vineyard resulted in wine that was different from a vineyard abutting it. In that way, the highly parceled system of vineyards and *climats*, the term in Burgundy for specific plots of vineyards, developed. And while styles have evolved over the centuries as a result of developments in winemaking, grape growing, and regulations, Pinot Noir and Chardonnay have been more or less synonymous with the place. In regions like Bordeaux, Cabernet Sauvignon, Cabernet Franc, and Merlot blends have dominated for hundreds of years. But climate change is occurring so quickly that it's not unthinkable that that could all shift, necessitating either new varieties that are more appropriate to the hotter summers and less-cold nights or that bud and flower at different times, which could help protect them from the most damaging effects of ill-timed extreme weather. The authorities have already begun that process for part of Bordeaux. If climate change forced a shift in the grape varieties planted in

Burgundy, Kuhlken told me, "That would upend the wine world . . . all of that knowledge stored for nothing."

In Texas, however, the issue isn't the wholesale change of varieties that are planted but rather on how to engage in the act of successfully farming grapes in the first place. One of the more insidious impacts of climate change is a shifting of pressures from various vineyard pests, which are being affected by the weather in ways that are causing real problems for grape growers. Pierce's disease, which I mentioned earlier, used to be relegated to more southerly climes in the Northern Hemisphere, but climate change is allowing the glassy-winged sharpshooter, which is the main spreader of the bacterium that causes PD, to travel farther north, exposing a wider range of vineyards to the disease-causing bacteria they transmit. Ron Yates had to replant almost all of his Sauvignon Blanc because of Pierce's disease in April 2021. And even before the February storm, he and his team went through their estate vineyard and spray-painted red hashes on the vines that they suspected had been compromised by PD. After the storm came and went, "Those vines went away," he told me. "Those vines did not come back." They had been weakened by PD and killed off by the storm: a climate-change one-two punch.

The unpredictability of the weather, and the changing of the climate, mean that an unusual collection of varieties have the potential to thrive there. Add that to the tendency in Texas to buck trends—Kuhlken contends that, though the state's cultural and political conservatism has gotten the most attention recently, it is, at its core, a place where libertarianism thrives most of all—and all of the ingredients are present for a wildly unique and exciting wine culture. "When Texans first started making wine in any seriousness, in the 1970s and well through the nineties and into the early twenty-first century, they did keep trying to make Cabernet Sauvignon, Merlot, Chardonnay, etcetera," she said. "And it is still being made," but less and less. Growers, she explained, began "realizing they're not thriving. . . . Why in the world are we trying to compete with Cabernet Sauvignon from California, right? We're not going to be Cali Cab; it's really, really darn good." So many top growers and producers made the switch from what were perceived as the most commercially viable varieties to the ones that, though perhaps less widely known in the larger world of consumers, would be better suited to the vagaries of the Texas climate. In other words, Texas would make its name with quality, not familiarity, which is how I was able, for example, to taste two different single-vineyard reds made from the Teroldego grape, a variety rarely seen outside of northeastern Italy. But if the examples I sampled in the barrel room at Pedernales are any indication, the variety could have a serious future in Texas: The wines were terrific.

It was a high mental hurdle for Texas growers and producers to get over, but once the best of them realized that they're likely never going to be able to make Cabs and Merlots that could compete with the greats of Napa and Bordeaux, and that Hill Country Pinot Noir was a fool's errand, a whole world of possibilities opened up. "Once you go off that beaten path," Kuhlken told me, "it's like, 'Well, why not? Let's just go see what the world of wine offers.' Because what you start to realize is, there's a lot of varieties that no one has ever heard of, and some of them are good fits for us." Another key factor is the fact that this is Texas wine, not iconic bottles from a region with a long and illustrious history. There's freedom, Kuhlken argued, to be found in that lack of expectation, in that lack of so-called snob appeal. In general, "People are going to stick up their noses if it's Texas wine," she said. "You're not going to convince them that you're going to be the best of whatever, so you might as well be at least eclectic and interesting about it because you can't be snobby and be making Texas wine. It's just not going to happen. And so you don't get stuck in a rut." That willingness to pivot, to not solely rely on the received wine wisdom of the ages or to espouse any sort of wine orthodoxy, means that Texas is uniquely well positioned to be able to navigate the ravages of climate change in ways that aren't necessarily the case in more venerated parts of the wine world with more calcified wine cultures. On an individual level, of course, grape growers and winemakers all over the classic growing regions of Europe are willing to experiment, but the regulations that dictate how and which grapes are grown and the ways in which they're vinified and aged are often slow to change; that's the nature of bureaucracy. As a consequence, they are generally less nimble in the face of the changing climate. They're not denying it or being caught flat-footed, of course—Bordeaux recently announced that a handful of new varieties would be permitted in part of the region, a nod in the direction of climate change, and producers all over the world are looking at the situation with open eyes and a deep willingness to change—but in Texas, with its lack of a long winemaking history, pivoting and experimenting can happen much more quickly.

Cultural considerations are important too. Texas, after all, despite its reputation for conservatism, is a far more individualistic place than it's often portrayed. "[P]eople just decide to do what they want to do," Kuhlken said. "And they're just going to go and do it . . . until someone tries to stop them. And then, *even* then, they're going to be like, 'Well, why are you trying to stop me? You're going to have to have some really good reason for that,' whereas I would say California is a more collectivist mentality." Later on, she added: "No one who wanted a corporate life went into the Texas wine industry. They went in to do what they thought was needed to be done. And I mean, there's no one who can really claim at this point like, 'Oh, I figured it out, you should all just follow my lead.' There was no one who can claim that, right? All of us have every right to say, 'Hey, I may have a better idea.'"

One person who had a better idea was Robert Young, MD, who still has the authoritative yet calming demeanor of a man who spent thirty-plus years as a physician. Today, as the co-owner with his wife, Brenda, of Bending Branch Winery, he goes by Dr. Bob, and while younger winemakers tend to get the lion's share of attention these days, he just may be the person to revolutionize winemaking in Texas and beyond, even though he's approaching his mid-seventies at this point.

"When I took a look at Texas fifteen years or so ago, I saw the red wines that were being produced, and they were not the kind of red wines that *I* wanted to produce. That was just a simple thing, really. It was just way too hot. Sure, you get plenty of sugar, you get a short harvest season and plenty of sugar . . . but you don't get phenolic maturity in some years," he explained. This is a common problem in hotter wine-growing regions. Wine grapes, in very broad terms, generally need around one hundred days to ripen into the kind of fruit that will result in wines with structure, complexity, and generosity.

On a technical level, that means that each individual berry achieves both sugar ripeness and phenolic ripeness (the alterations in the tannins and anthocyanins present in the grape's skin, stems, and seeds that allow the fermented wine to attain the desired level of complexity without being overwhelmed by the astringency that is typical of immature tannins). Optimally, both sugar and phenolic ripeness are reached at the same time, or close to it—this balance is one of the key factors guiding picking decisions in a vineyard. But in hot-climate regions like Texas Hill Country, a short growing season means that while the grapes regularly develop enough sugar, phenolic ripeness often lags behind. And waiting for all polyphenols and tannins to further develop risks the grapes becoming overripe, which results in flat wines that lack enough acidity. . . but you still need that phenolic maturity. "If you don't have those phenols, those polyphenols, the tannins and the anthocyanins, then it's pretty difficult to make great red wines," Young explained. "You won't get the flavors, the body, or the structure, you won't get longevity and all those kinds of things, and that's not the wine I wanted to make," which is where Young's training as a doctor and his unwavering faith in the power of science and technology resulted in his proverbial a-ha moment. "I said, 'Well, I can't change the environment, but maybe I can change the way [I] make wine, and address this problem.'" He asked himself a deceptively simple question that has had ramifications far beyond his estate vineyard in Comfort, Texas: "'How do I, as a winemaker, deal with this problem and figure out how to extract more of those polyphenols from the Texas red grapes?'"

So he started digging into the research. Pretty quickly, he realized that only around 25 to 40 percent of the polyphenols in each grape are typically extracted during the winemaking process. The rest of it ends up in the pomace, the paste of crushed skins, seeds, and stems that's left over after the wine has been made. "There's *stuff* in there," Young explained, referring to the polyphenols and anthocyanins he was after. "It's just *stuck* in there. And different types of [chemical] bonds that are hard to break—not all of them break during a typical alcoholic fermentation. So I focused on methods of how to extract more of those polyphenols from those grapes since we often didn't make enough of them, at least in my opinion."

Young had completed the online winemaking and oenology certification program from the University of California, Davis, and while he was going through it, researching methods of polyphenol extraction, he found an obscure research paper that had been published in France around twenty years earlier. In it, the authors describe freezing Cabernet Sauvignon and Merlot, thawing them, and then proceeding with a normal fermentation. "The result was they extracted about 50 percent more polyphenols, about 50 percent more color, and about 50 percent more tannins just by freezing it in advance," he said. "A lightbulb went on in my mind: Why is nobody doing this, particularly in hot-climate regions? I just started doing it. . . . I have a lot of science in my background. I like to experiment. I said, 'Hey, we're going to try this out.'"

The process is called "cryo-maceration," and his first attempt involved freezing his grapes for two to three months before thawing them out and initiating fermentation. It worked better than he could have imagined: "It turned out to be the first 100 percent Texas Cab to win a double gold in San Francisco International," the prestigious wine and spirits competition. "I said, 'Okay. I think we're on to something here maybe.' The next year, I did it with our estate Tannat, and it got the Top Texas Wine award at the Houston [Livestock Show and Rodeo] International Wine Competition."

For all of his success, however, he still wasn't able to extract as much color from the grape skins as he wanted. And if he could extract even more of the polyphenols—his cryo-maceration was pulling around 20 percent more color and 50 percent more tannins than standard fermentation—so much the better. This is when he decided to try a technology called flash détente, which involves quickly heating and then cooling the grapes in a vacuum. "When you weaken the cell walls with the heat and then you cool in a vacuum, the cells just open up. Those polyphenols that are stuck in the skins or inside the cells, many of them are freed up," Young explained. "I did about five years of data collection on this. I sent the resulting wines out to California for testing, and on average, [I] get a 100 percent increase in extraction of polyphenols from that. It works. So to me as a winemaker, if you want to make that style of wine, if you don't want to make . . . light reds, this is a solution to the problem." And in a region where climate change is challenging seemingly every decision a winemaker or vine grower faces, it could prove to be a remarkably useful tool. Texas winemakers, he predicts, are "going to need it more and more. As the climate gets hotter and drier, as is predicted by academics here in Texas . . . the season is going to get shorter and shorter, and there's going to be less phenolic ripeness in the fruit in many areas of the state." And as storms and unpredictable weather keep roiling the region, often right around harvest time, these techniques could allow grapes to be picked a bit earlier than they optimally would and still afford the winemaker the opportunity to craft the kind of wines he or she wants to. I've tasted a number of Young's wines, and the proof is right there in the glass: They're expressive, generous, and in many cases, absolutely delicious.

Younger winemakers may get most of the buzz, but sometimes, it's a seventy-something former doctor who has the vision and the moxie to change everything.

One of the most vexing issues facing grape growers and winemakers around the world is the unpredictability of climate change. In a place like Texas, which has always dealt with variable weather, the extremes are getting worse, and the old assumptions about how they'll impact one place as opposed to another are getting blown up. In 2017, for example, the predicted path of Hurricane Harvey sent jolts of fear through the High Plains wine community—it was late August as the storm approached, right around harvest time (the warmer Hill Country growers had mostly brought in their fruit by that point), and the threat from massive amounts of water being dumped would have meant diluted wines, shattered grapes, flooded vineyards, and more. What actually happened was that the "swirl of air," as Julie Kuhlken called it, pulled down cool, dry air from the Midwest. "And so during a period where normally it would be hot and dry potentially, and you're just trying to harvest everything at once because everything's ripening all at once, or you're dodging storms because it's also one of those transition periods, so you're constantly like, 'Why is it going to rain tomorrow? So I should harvest today?' None of that happened. Instead, we had almost California-like conditions, where it was cool and dry and consistent, and we could pick each variety just where we wanted to rather than just say, 'Okay, emergency dictates that I pick today and now.' And yes, you could spread out the harvest in a way that was just perfect for wine quality and phenolic ripeness and acidity and the tannins—everything that you want to be attentive to."

Meanwhile, the category 4 hurricane "dropped more than 50 inches of rain in some sections of the Houston area, resulting in $125 billion in damage and directly or indirectly in more than 100 deaths," according to the University of Houston.[5] The stalling of the system that pulled down all of that cool, dry air from the Midwest and led to such a banner harvest in the Texas High Plains utterly destroyed the lives of others. That's one of the main issues with climate change in Texas and so many other parts of the world: its unpredictability and extreme consequences. For now, however, the most forward-thinking winemakers in Texas are doing everything they can to leverage the good, mitigate the bad, and carve their own path forward, even in these thoroughly disorienting times, which come to think of it, is what Texans have always prided themselves on doing.

"We've been a working laboratory, if you will, for effects of the climate on various varieties," Young noted. He added, "I think you've already learned one of the key things is the diversity of the impacts and the unpredictability of the impacts of climate change in the state. You get all these different things. We've got heat waves; we've got droughts. Several years ago, we had a hundred-year drought. We also, a few years ago, had a hundred-year heat wave." He went on: "We had fifteen days in a row over 105, as an example. We get all these hailstorms now that pop up unpredictably multiple times throughout the year. More spring frost. We even get some flooding. Newsom Vineyards, which is one of the vineyards that we buy a lot of fruit from, is one of the oldest family-owned vineyards in the state of Texas, [and] they had flooding in their vineyard about two and a half weeks before harvest. There was several inches of water standing on the vineyard floor, which is crazy. We even had, in past years, a touch of the forest fires. It's just not been that much. So the unpredictability is really there, and we have seen historically since 1970 an increase on average temperatures by about two degrees Fahrenheit since 1970 to 2020." Yet the wine from Texas keeps on getting better and better. Dr. Bob, Julie Kuhlken, Ron Yates, Dave Reilly, and the rest are all finding ways to make it work. And while they and their colleagues continue to navigate a path into the future, the rest of the wine world may likely—and unexpectedly—begin looking to Texas, of all places, for some indication of what a way forward might look like in this strange new climatic world.

8

THE PHILOSOPHER-FARMER OF THE WESTERN CAPE

It's a sad fact that natural disasters tend to have the biggest effect on the poorest and least powerful in any given society. Sure, hurricanes can wipe out entire neighborhoods regardless of the average family income in that particular zip code, and wildfires burn through countless acres whether the land is worked by a hand-to-mouth farmer or owned by a gentleman vintner who made his fortunes in the financial markets on the other coast of the country: Mother Nature doesn't discriminate. But the wealthy and powerful are not only able to take often more proactive measures in the face of impending catastrophe—there were reports about some of the more well-off residents of New Orleans, for example, leaving their homes for the safety of higher-ground hotels as Hurricane Katrina's ferocity became apparent back in 2005—but the places they live tend to have better infrastructure to begin with. Their neighborhoods and towns are less likely to lose power in the first place and are generally located in areas that are less prone to the worst impacts of what the natural world throws at them.

But what happens when the biggest natural disaster of them all—climate change—threatens an industry that has become an important stepping stone for a population of people in a country whose official policy for decades was to make sure they never had those opportunities in the first place?

This is the unfortunate situation that South Africa finds itself in today.

The National Party came to power in 1948 and "its all-white government immediately began enforcing existing policies of racial segregation," explained the editors of history.com[1]. The enactment in 1950 of the Population Registration Act then "required people to be identified and registered from birth as one of four distinct racial groups: White, Coloured, Bantu (Black African), and other," noted South African politician and activist Helen Suzman. "It was one of the 'pillars' of Apartheid. Race was reflected in the individual's Identity Number."[2] South Africa's apartheid government worked to maintain what it hoped would be a permanent underclass, forcing non-whites to live in sprawling townships whose physical infrastructure was often jerry-rigged by the local population and whose educational opportunities were severely limited, to say the least. Even today, the drive from Cape Town to the winelands of the Western Cape take you past vast, almost horizonless slums of ramshackle houses, their corrugated roofs slapping in the wind and the labyrinthine streets unfathomable in their complexity. Nelson Mandela was released in 1990 and he and the African National Congress took the reins of power in 1994, but decades and decades of official and de facto policy resulted in generations of people who in too many cases lacked sufficient education and opportunity . . . or much of anything else.

Yet the wine industry has proven to be an important and unexpected stepping-stone for many people whose parents and grandparents were forcibly denied access to that world and to the opportunities it presented. During my first visit to South Africa, in 2014, I was struck by the number of young Black sommeliers, tasting-room managers, winemakers, and more, and subsequent visits have seemed to show that their ranks are growing. A look at the numbers bears that out: in 2000, Carmen Stevens became the first Coloured woman to hold the title of winemaker at a major wine group, crafting the wines at the Welmoed winery for Stellenbosch Vineyards, and in 2004, Ntsiki Biyela, at the Stellekaya winery, became the first Black woman winemaker in the country; she now helms her own brand,

Aslina. According to the South African newspaper *Mail & Guardian*, "There's since been several Black winemakers who've shaken up the 355-year-old white industry. There's Seven Sisters, founded by Vivian Kleynhan, who makes seven wines, each dedicated and named after her seven sisters, who are also involved in the business of winemaking. Other Black-owned wines that have entered the space include Ses' Fikile Wines, Lithathi Wines, M'Hudi Wines, Thandi Wines, House of Mandela, Adama Wines, Bayede!, Imvula Wine, the . . . House of BNG and others."[3]

(Before continuing, however, a note on the language I'll be using in this chapter is important: I will be employing the term "Coloured" not because it's part of my personal lexicon, but because it is very much a term that is used in South Africa even today, and generally without the same cultural and linguistic baggage that it has here in the United States. Its use goes back to the 1800s, and according to Jim Clarke in *The Wines of South Africa*, "Coloured is the accepted name for the mixed-race group prominent in the Western Cape. This group is descended from local Black people resident at the time of European colonization, Khoisan peoples most especially; European settlers; and slaves, in particular those from Southeast Asia. The term does not carry the same negative connotations associated with it in the U.S. and elsewhere."[4] Its awful roots—as a way of defining and classifying people and as a pejorative during the years of apartheid—cannot and should not be ignored, but the fact remains that it's a term of description in South Africa that is used by all members of society, regardless of racial or ethnic background.)

Still, major strides are needed. According to a 2019 article in the trade publication *SevenFifty Daily*, "Approximately 60 percent of the country's 300,000-person wine industry workforce consists of employees from previously disadvantaged Black and mixed-race groups, according to research conducted by the Western Cape Government in Cape Town. However, while they comprise more than half of the wine industry, these workers aren't likely to be offered valued positions or leadership roles—and few have any business ownership stake or land equity." Inevitably, simmering tensions occasionally boil to the surface. The article continued: "[T]he May 2018 takeover of a Stellenbosch vineyard owned by Stefan Smit—a white farmer—by Black residents from the nearby Kayamandi township exemplifies the racial tensions that still divide the country, even decades after the end of apartheid."[5]

The situation seems to be improving, however, with socially conscientious farmers, landowners, and proprietors of wineries working to become not just more inclusive but also more thoughtful in the nature of those changes. Thokozani wines, for example, allows its workers, more than two-thirds of whom are Black or Coloured, to acquire equity stakes of the company through salary buybacks. "The brand also gives equity in buildings and land assets, which means that shareholders have additional revenue streams, tied not only to wine sales," the article in *SevenFifty Daily* continued. "Dividends were paid out the last three years, rewarding the shareholders' efforts."

The wine industry around the world has, in recent years, increasingly focused on inclusion, on providing the kind of unique opportunities that the making, marketing, and selling of wine is full of, to people who for far too long were cut out of it. There's a reason, after all, that wine was historically perceived to be the provenance of moneyed white men—because to a great extent, and with a few notable exceptions, it was.

In South Africa, that focus on inclusion has an even deeper meaning, and the headway that Black and Coloured people there have made in the wine world is, while still a work in progress, positive. The problem is that just as the industry there is hitting a new peak of international respect and renown, benefiting a wider swath of society than ever before, climate change is threatening it in ways that are frankly terrifying, shaking the very structure underpinning the strides that the victims of apartheid have made within the world of wine.

As always, it's the least advantaged who are at the greatest risk.

According to a 2012 article in the trade publication *VinIntell*, "South Africa falls within a vulnerable region as far as climatic change is concerned due to its geographical location and its low level of coping capacity. Furthermore, South Africa is one of the highest emitters of GHGs [greenhouse gases] in the world, ranked 19th in 2005 if emissions from land-use change and forestry are excluded. Of all the sectors, the agricultural sector has been identified as being particularly vulnerable to climate change impacts."[6] The authors continued:

"Based on scientific findings, the South African Fruit and Wine Initiative says the most prominent biophysical impacts of climate change on the South African agricultural sector include a decrease in water availability, a shift in seasonal temperatures and climatic patterns, and an increase in the prevalence of pests and diseases." They added that "the impact is five-pronged[,] namely change to regional climatic patterns, changes in the distribution of pests and diseases, in energy and fuel prices, increases in market pressures and the impact of carbon pricing." Add to that perhaps the most painful of all: a potentially devastating impact on the people who can least afford it.

This, then, is the story of one man who is bringing together the fight for equality, dignity, and the environment in unique and remarkable ways.

When Johan Reyneke was a young man, just out of university and starting his postgraduate work in environmental ethics and philosophy, he found himself in a situation that's familiar to grad students all over the world: he needed more money than his student stipends provided. Yet because Reyneke has always been the kind of person who thinks deeply about his personal actions and their effects on the world around him, he didn't apply for work at the local café, or spend his free time shelving books at the university library, or any of the other more typical grad student jobs. Instead, he began working in a field that was totally unexpected of a highly educated, upwardly mobile young white man in 1990s South Africa. And the time he spent doing it has, all these years later, unexpectedly impacted a broad swath of the nature and texture of the wine industry in South Africa.

"I started working as a farm worker or a laborer, a vineyard worker," he told me this past winter over Zoom. It was a steely day in my Philadelphia suburb, one of those early-winter mornings when the radiator in my old house could do little to keep up with the encroaching chill. The latest COVID spike had made travel more or less impossible, and I was staring down another pandemic winter at home. On my computer screen, however, the weather was perfect, a bright, full-of-promise summertime day in South Africa, and the birds outside Reyneke's window, though I couldn't see them, provided the background soundtrack for our conversation. I'd soon learn that those birds are only a small fraction of the life his ninety acres of land teems with; his family farm in the Polkadraai Hills, just outside of Stellenbosch, was the first to be Demeter certified as biodynamic in South Africa. As such, the land is home to a mind-boggling range of wild and domesticated animals, insect life, plant diversity, and more. He's not the kind of guy who holds back in conversation—Reyneke is passionate about what he does and follows the trail of his ideas to their logical and often surprising conclusions—and he jumped right into explaining why he decided on not just an intensely challenging, physically exhausting way of earning some extra cash as a student but also on why it was so important to him that he do so.

"I haven't really worked in the Napa and Sonoma vineyards," he said, "but I can imagine that the bulk of the physical work would probably be done [by] people from Mexico." He was right; it is. "In South Africa, the bulk of the physical work was done by the Indigenous peoples, people of color, people who had been excluded from the sort of mainstream economy, especially because of apartheid." The hard work of planting and maintaining vineyards and farms was very much *not* something that white South Africans like him generally did. In that regard, it mirrored the historical situation in wine regions around the world; there are exceptions, of course, and things are mercifully changing, but even in Europe today, it's not uncommon to see roving teams of vineyard workers from the poorer parts of the EU toiling away between rows of vines. So when Reyneke decided to earn the extra money he needed in the sun and the soil, it was a personal decision as much as a political statement. Yet his energy and idealism at the time—neither of which, I quickly learned, have diminished over the decades—led him to want to work alongside the people who had been shunned by white society, to understand their labor and at least a part of their lives in a way that he had never been able to before.

"I was . . . quite young and idealistic, you know? I was studying philosophy. I was in my twenties, thinking about stuff and asking a lot of questions. I think it was a combined effect." He read widely in the evenings after his long days in the fields and developed a particular affection for a handful of key American thinkers. Thoreau's *Walden* had a deep impact on him, and Aldo Leopold's *A Sand County Almanac* proved to be formative. Leopold's tome, though less widely read than Thoreau's most famous work, has been a defining text for people like Reyneke since it was first published in 1949. In it, Leopold argues, according to the Aldo Leopold Foundation, "Ethics direct all members of a community to treat one another with respect. [And] a land ethic, Leopold wrote, 'simply enlarges the boundaries of the community' to include not only humans, but also soils, waters, plants, and animals—or what Leopold called 'the land.'"[7] Reyneke took that idea and ran with it. "I think with knowledge comes obligation to act," he told me. "It's very difficult to read [those ideas] and just forget about them the next day. In my studies, I kind of read a lot about nature and our interaction with nature, and the interaction was a very specific part because South Africa, in a way, is almost like a microcosm of the world. It's like we're the first and third world in one. A lot of the dynamics that one would see playing [out] on a global scale, we see playing out with a domestic version. . . . There's no real border control or anything between the first and the third world: you can literally walk from a mansion to a shack and from a shack to a mansion in ten, fifteen minutes in the same country." And the people living in those two thoroughly divergent worlds may never meet, may never interact.

That dichotomy struck him as particularly galling, and it was magnified by all sorts of cultural and perceptual hangovers from the recently ended apartheid. It all came to a head one day when he read an article about a poor man in the east of the country who, he recalled, "struggled to eke out a living, didn't have land that he owned," Reyneke recounted, telling the story as if it were a fable. (Note: I haven't been able to independently corroborate this.) This man, however, managed to secure "a small portion in this [UNESCO-protected] wilderness where he found a little space to live. And he grew some of his vegetables, grew some food for himself to eat." One day, Reyneke told me, a hippopotamus ambled by and ate all of his food. In a moment of rage, the man shot and killed the hippo. Shortly after, he was arrested. It seemed like a fairly straightforward case of a man illegally farming land that wasn't his to cultivate and then killing an animal that's indigenous to that very land. For most people—and especially for many well-to-do white people—it all added up to a relatively straightforward case. But nothing in South Africa, especially in the aftermath of apartheid, was all that straightforward. "A lot of people freaked out," Reyneke told me, "because how do you go and grow stuff in a UNESCO World Heritage site, and then you have the audacity to go and shoot the wild animals that live there?" Not everyone saw it that way, however. "The other half of the people got really upset," he continued, "because some people seemed to be more concerned with the well-being of the hippopotamus than of the fellow human being. It kind of brought this whole environmental development thing to a head, and I was really trying to find practical solutions for that as well." In this one tragic event—for the man, for the hippo, for the society that made a situation like that even possible or necessary in the first place—everything that Reyneke had been reading about, and pondering during his hours of labor in the fields, seemed to come to a head. "Although my thoughts in working in the vineyards were all things nature and environmental, that definitely [had] a social dynamic [aspect] from day one as well."

As he was reading these books and as South Africa was engaging in the kind of post-traumatic reassessment of what it meant to be part of a newly transformed society—at least in official terms, if not yet for everyone in their own daily reality—several important realizations occurred to him as a field worker. Among the most transformative was the very nature of the chemicals he and his colleagues were being forced to use. "I was part of a team of roving laborers to work on different farms . . . and we were [given] herbicides and pesticides and fungicides on a daily basis to work with," he explained. "And it kind of sucked because on the can there would be a skull and crossbones, and it would tell you that it could cause cancer, and that you had to wear protective clothing, and you weren't allowed to eat or smoke or anything unless you washed your hands thoroughly. I just thought, this is such a crap product to have to work with every day."

He eventually fell into a routine; once he had finished his work for the day and was off the farms for the evening—body thoroughly scrubbed from whatever poisons he had dumped into the land that day in an effort to make it more fertile, more productive—his mind would continue to race. There was a cresting feedback loop between his visceral experiences in the fields and his intellectual ones at night. "My physical experience there sucked, and then in the evening, I could see how we were just encroaching upon nature," he said. "And where we used to be vulnerable, and nature was something we were supposed to protect ourselves from, the whole dynamic changed, and nature became vulnerable, and *we* became the aggressor. And one almost had to protect nature from ourselves, and from our worldview."

This marked a sea change in how Reyneke saw not just the act of modern farming itself but also in his understanding of how the impact of that work affected certain populations in far more dangerous and destructive ways than others. His experiences as a roving farm laborer—work that he voluntarily chose to engage in, as opposed to many of the people toiling alongside him, for whom that kind of work wasn't a choice but a last resort—were the bud of what would ultimately blossom to grow into one of South Africa's most important and environmentally conscientious grape-growing and winemaking operations.

"I didn't set off to do [these] things," Reyneke told me. "I set off to farm in a more environmentally friendly way, and that led me down a path of organics and biodynamics. It was just coincidental. I didn't set out to become a good organic farmer or a biodynamic farmer. It was just [that] I was looking for help; I was looking for people who could help me to farm without using all these aggressive and highly toxic and hazardous chemicals in my daily doings." And with wine, given the nature of the product and the culture around it, all of this was thrown into even sharper relief.

Over the course of more than fifteen years in this field and having traveled to dozens of countries to more clearly understand their wines, spirits, food, and culture, I've learned that wine professionals, in general, tend to be both deeply and widely read and to look at wine as infinitely more than just the liquid sloshing around in the glass. Reyneke embodies this outlook. He told me that at the same time he was learning more about growing grapes and making wine, "I started getting a very weird feeling about the wine industry in general because beauty cannot come from ugly, goodness cannot come from badness. You know why? It's such a beautiful product, and it's such an amazing thing. I mean, if you think how we use wine to celebrate and to toast and to basically just enjoy [life]. It's a sophisticated beverage with incredible cultural backgrounds." So then, he continued, you ask yourself, "Where does this come from? It comes from the vineyard, and a lot of things happen in the vineyard. The aggression—in terms of how we work with the natural environment in the vineyard—that didn't sit well with me in that context. Also, the almost abuse or exploitation of the workers [weighed on me] as well. I just wanted to stop that and have a completely different approach to working with people and to working with nature," which is where the Cornerstone Project (more on that later) and biodynamics came in.

For Reyneke, owning a vineyard and winery wasn't just a way to become the kind of steward of the land that he dreamed of being but also the kind of member of society that he desired to be. "It came down to a point where I tried to get my colleagues to start a wine business with me," he recalled, "and they declined because the odds were against us." No wonder: "Rich people start a wine company and lose a fortune," he continued, paraphrasing an old wine-business joke. (How do you make a small fortune owning a winery? Answer: Start with a big fortune.) "We were a bunch of poor people trying to start a wine company and become rich. Also, I had zero official . . . education in commerce or viticulture or oenology. You know, a degree in philosophy can be fun, but it's not a guarantee in terms of financial success down the road. And most of my colleagues [from the farm-working days] were illiterate because they had to leave school when they were ten or eleven years old to also work in the fields to help their parents put some food on the table. It wasn't a great start other than the values and the intentions, which were present." In other words, they knew the value of hard work—they just didn't have much opportunity to leverage them to something greater.

It wasn't easy even after he raised enough money to buy the land and get started with planting . . . and indeed, for some time after that. Reyneke understood that he wouldn't be able to have the social and environmental impacts that he desired if the vineyard and farm didn't become successful: no matter how good his intentions were, he'd accomplish nothing without money. He was trying to merge his two passions—the health of the land and the thriving of previously subjugated people—and the way things started were, as he told me, a complete disaster. "From an environmental point of view, I had every pest, bug, and disease you can find in a vineyard, in *my* vineyard, within six months. It was a nightmare. I was loving Mother Nature, and she wasn't loving me back. I didn't know what to do or where to go." Word got out that he was struggling, and he found himself being introduced to all kinds of vineyard and farming experts, including Jeanne Malherbe, whose farm's name, Bloublommetjieskloof, translates from the Afrikaans to Blue Flower Valley. He affectionally calls her "a doyenne of organics and biodynamics," he said. "Not in viticulture—she didn't even drink wine." Her expertise, rather, was in harmoniously farming vegetables, flowers, and other plants and crops in a way that actually benefited the environment. "She explained to me that I was organic by neglect, and I had to become so by design."

With that, he continued, "She planted a few crucial seeds in my mind. I think [one of the most important ones] was that farmers use herbicides and pesticides and fungicides, not because they want to, but because they *have* to. If you're going to stop using those things, your farm will fall apart unless you put an alternative methodology in place to address those specific issues," which seems easy enough in theory, but one of the main reasons that so many farms and vineyards were historically reluctant to convert to organics and biodynamics in the first place is that once you take away one or two of the chemical inputs and try to replace them with nonsynthetic ones, the proverbial dominoes may start to fall, and there's no way—or little way—of telling what might happen once that process begins. Reyneke, however, was determined and broke the steps he'd need to take into what he called bite-sized chunks.

For many grape growers who move to organics in the vineyard, one of the first steps is to replace the use of herbicides and fungicides with a spray composed of copper sulfate and lime, which according to an article in *Wine Spectator*, have been widely used since the 1880s against bacteria, fungus, and the dreaded downy mildew that grape growers around the world fear. "But," the article notes, "risk assessments by public authorities like the European Food Safety Authority (EFSA) show that copper compounds pose risks for farm workers, birds, mammals, ground water, soil organisms and earthworms. These risks make copper unpalatable to many vintners."[8]

Reyneke knew this and didn't want to replace potentially dangerous synthetics with potentially dangerous copper just because it was on the list of approved products for organic viticulture. "I didn't really want to spray copper, even though it was organically the appropriate thing to do," he explained, "because it's a metal, and it builds up in the soil and becomes a self-fulfilling exercise that will eventually start killing off the microbial life in the soil that you're trying to build and to foster." So instead of accepting the received wisdom of organic regulations without question, he decided to focus on the soil itself as well as the entire ecology that he was working with on his particular plot of land. Specifically, he told me in an email, he began using trichoderma, a fungus "that acts as a predator for downy mildew. We also use their metabolite (excreted by trichoderma) to increase the brix (sugar) levels of the vines, which in turn increases their natural resistance against downy mildew."

This outside-the-box thinking became something of a calling card for Reyneke. Snails, for example, were an early problem in his vineyard, but instead of using snail bait, which can be poisonous for dogs, cats, and wild birds, he brought in ducks to roam the land. "They will eat all the snails, and they will crap everywhere and give you a lot of free fertilizer," Reyneke bluntly explained. As for leafroll virus, which is spread through the insect equivalent of saliva from the mealybug, "I used to have to spray a very toxic pesticide [during my roving farm laborer days] that was actually banned for use in Europe back in the day." Ultimately, he told me later on, it's "about bigger yields, greed, and letting nature (and future generations) pick up the tab." On his own land, however, he found a different and far more healthy approach. Again, however, this one was counterintuitive to him at that time and may have been perplexing to many other farmers and gardeners around the world. "We understood that all we had to do was allow some dandelions to grow in our vineyards because the root of the dandelion was the preferred habitat of the insect and if you just allowed some to grow, the mealybug would actually move out of the vine and not be a problem." He also explained to me that, "Over time, we've also had a huge build-up of natural predators like wasps and lady bugs that predate on the mealybugs also." The use of more natural methods typically has long-term benefits. "If you go and you spray an herbicide, you think you're killing off weeds and things. But in effect, you're [also] decimating the habitat of a gazillion things.

The only place you leave for them to go is the vines." And if they find their way to the vines, you need to spray. It very quickly becomes a vicious cycle.

Received wisdom has it that successfully growing any crop requires efforts to be made to ensure that as many of the soil's resources go to that particular plant as possible—weeds, in other words, are the enemy. "If you go outside and you want to grow stuff in your garden, you want to grow flowers or you want to grow veggies or whatever," Reyneke said, "the first thing you're going to do is to remove all the weeds and the things that grow there because you want all the love to go to your plants." But healthy soil actually requires weeds, he explained, just as much as it needs all of the other flora that grow in it, and the animals that walk on it and pass waste into it, and the insects that live inside it. Healthy soil teems with life that's a result of all the many things that live and grow in and on top of it. That meant that Reyneke had to decide between the health of his soil and the growth of his vines, which seemed at first to be in opposition to one another. "That created a paradox," he recalled. "What was I supposed to do? Was I supposed to remove the weeds and the things for the sake of the vines, or was I supposed to leave them for the sake of the soil? Then I got [some] very interesting advice." A professor from Geisenheim University in Germany visited Reyneke in South Africa, and he explained to him that the issues he was grappling with were really only paradoxical in a short-term view. "'The longer-term view you take,'" the professor told Reyneke, "'it becomes less of a paradox. And eventually, inversely, it becomes a synergy of sorts.'" What at first sounded like farming advice from some sort of Germanic Yoda turned into one of the linchpins of change for Reyneke; he realized that the health of the soil and the health of the plants that grow in it were inextricably linked, and he could nurture both in concert. All it required was altering his view of the land and a healthy dose of patience.

Reyneke decided to reframe his work; instead of being a farmer in the traditional sense, he would become a soil farmer. He of course continued to grow grapes and other crops, but he became increasingly obsessed with the idea of building the soil, of increasing the amount of humus in it, of organic matter, which isn't only nutritionally beneficial to the plants that grow in it, but it also retains moisture and reduces runoff. The impacts are tremendous, especially in a world that's being more and more impacted by the unpredictable weather that results from climate change. "If you could increase the humus levels in your soil up to 5 percent," Reyneke explained, "the resilience of the plants that live there would increase, or could increase, by as much as 300 percent. That, for us, was particularly significant because we're farming some of the oldest and most extensively weathered soils you can find anywhere in the world. They're much older than your soils in California or the ones in France or Italy or in Europe." And with climate change becoming an ever-more-pressing matter, he knew that this was the right move.

So he set about building his soils, coplanting grains and legumes, reintroducing the wildlife and so-called weeds that more conventional farms and vineyards would work to eliminate. Eventually, analyses of his soil showed that his numbers were indeed increasing and his "untidy" vineyard, as he called it, was actually teeming with life, both above the ground and beneath it.

Little did he know that he was preparing his land to survive in a rapidly changing natural world.

During a 2017 visit to South Africa, I found that traveling around the country was no different from touring through any other—climate change was a topic of conversation in winery tasting rooms and on walks through the vineyards, but aside from that, it didn't seem overly pressing among people who weren't directly dependent on the land for their living, at least in casual conversation. In 2018, however, that sense of comfort was flipped on its head in dramatic fashion: water shortages had become so severe that not only was the agricultural sector being affected but the lives of everyday South Africans were too—and in startling ways. Colleagues who visited that year peppered my inbox and phone with stories of hotels putting up signs asking guests not to shower every day to preserve water. Climate change, and the resulting water rationing, was now impacting the all-important tourism sector. For a country so reliant on tourism, and on high-end tourism at that, this was a major sign.

On January 1, 2018, the government of Cape Town announced what were called Level 6 water restrictions: no more than eighty-seven liters of water per day were allowed to be used by each person.[9] By February, that number was lowered to fifty liters. For a sense of context, showerheads with the WaterSense label use no more than two gallons per minute, or just over nine liters. Standard models, according to the United States Environmental Protection Agency, use around two and a half gallons per minute, or more than eleven liters.[10] For a ten-minute shower, that's still approximately ninety liters of water even with efficient heads, and over a hundred and ten liters for standard ones. A single shower, even one with a modern and efficient setup, would use up more than the daily allotment . . . and that's before any other use of water throughout the day, from washing a load of dishes to flushing a toilet to simply filling a glass to slake a summertime thirst.

It was all an effort to stave off what was known as "Day Zero." According to an article at Bloomberg.com, "In January 2018, when officials in Cape Town announced that the city of 4 million people was three months away from running out of municipal water, the world was stunned. Labelled 'Day Zero' by local officials and brought on by three consecutive years of anemic rainfall, April 12, 2018, was to be the date of the largest drought-induced municipal water failure in modern history." The article continued: "Photos of parched-earth dams and residents lining up to collect spring water splashed across news sites. The city's contingency plan called for the entire population to collect its water—a maximum of a two-minute-shower's-worth a day per person—from 200 centralized water centers, each serving the population equivalent of an MLS soccer stadium."[11] Fortunately, the mitigation efforts worked, and Day Zero was avoided, but the effects were felt widely and intimately.

"We were literally standing in buckets when we showered, and you were only allowed to shower for a minute or two," Reyneke marveled. "And then you had to use this bucket of shower water to go and flush your loo with. If you just did a number one in the loo, you would just let it lie there. If it was a number two, you would rinse it. It was an absurd experience, man." And for farmers like him, it was also felt in the land. "In that absurdness, it was almost impossible to build the soil because the farmers were struggling in the heat and in the drought and were hanging by a thread. We really had to be aggressive in our weed management and [did] everything we could just to pull off sustainable yields. We couldn't build the soil then."

Still, because he had been building that soil for so long before the 2018 crisis and because his vines and other plants had grown accustomed to living without any chemical inputs or an overabundance of irrigation, they survived. It wasn't a period of growth, of thriving, but their innate heartiness allowed them to pull through until the next year.

The rains mercifully came in 2019 and 2020, which allowed Reyneke to push ahead with his soil-building work in the vineyards. As a result, he told me, he was able to move away from weeds to focus more on companion planting—growing grains and legumes and grasses between his rows of vines to not just increase the amount of nitrogen in the soil, as well as the overall organic matter, but also to give the soil greater structure. This, Reyneke had learned, was growing increasingly important; one of the most common impacts of climate change that grape and grain growers and winemakers and distillers around the world have consistently referenced throughout my time reporting for this book is a change in how the rain falls. "If you look at global warming—and global warming is very real—that shower-in-the-bucket stuff wasn't a figment of my imagination. It sucked proper," he told me. But that's only half the story, because the other side of that coin is *too* much rain in too short a period of time. "Even the rain that we have now is not the rain that I grew up with as a kid," he went on. "We had a fine drizzling rain" as often as not, he recalled, but now, "[i]t's torrential stuff . . . it comes away and washes away your topsoil, and you have to really do more work to try and reduce the erosion." Breaking the drought of 2018 was an inherently good thing, but more intense rain over shorter spans of time "brings its own set of problems," Reyneke lamented.

Farmers like him are in a constant tug of war with a dramatically and rapidly changing natural world. Yet for all the struggles he has faced and for all the obstacles he has bumped up against, his land and the vines and other plants that grow in it are still better positioned to survive in the more challenging years than those of his colleagues who farm more conventionally.

Perhaps it was inevitable that someone like Reyneke would become such an important figure in both the social and environmental revolutions in the world of South African wine. Of course, he would likely tell you that there really is no separating them: they're two sides of the same proverbial coin. No one, I suspect, can read Leopold and Thoreau as closely as Reyneke did, while working on a roving crew of farm laborers, and *not* see the connections between the abuse of the environment and the mistreatment of the people working in it. So once he acquired the land that would become the Reyneke estate vineyard, farm, and de facto nature reserve (it had been a farm since 1863 and named Uitzicht, or "View," in Dutch; its north, south, and east flanks offer vistas of Stellenbosch and False Bay, among other wonders of the Western Cape), he knew that it would be much more than a place to grow the grapes that he would turn into wine. Instead, it would become a grand experiment in the ways in which an often damaging industry, as it was widely practiced at the time, could be both beneficial to the earth *and* to the vineyard and farmworkers whose well-being, despite their crucial work on the land, had never really been considered during the horrific years of apartheid . . . and in many instances, even after it was dismantled, as his experience as a laborer helped show him.

"We started the Cornerstone Project, and the only idea behind it really was that after Mandela was released, and it was the end of apartheid, that people were politically free and ideologically free, but money is a funny thing," he told me. "It's multigenerational. So [Black and Coloured people] were structurally excluded from the mainstream economy. And even though they were free . . . they were still poor, unemployed, and they didn't have places in which to live." The end of apartheid, the election of Nelson Mandela, the growth of a more representative government—all of these were necessary steps, but that's just what they were: steps. After nearly half a century of an official system that deprived non-white South Africans of not just freedom and dignity but also of adequate education and opportunity, poverty had become endemic in so wide a swath of the Black and Coloured population that it will take more than the vote, more than representation in government and growing opportunities, to climb out of it. It's a lament that I heard throughout my travels in South Africa, and each of the three times I visited, at some point I'd find myself in conversation with people across the racial spectrum about the long-term effects of apartheid, many of which can still be felt today.

There are amazing organizations throughout the country that aim to help lift people out of the cycle of poverty—and to give them the tools to more efficiently help themselves—and the wine industry has grown to become one of the more important. For Reyneke, as he continued down the path of transforming his land to a more natural state—or rather, to a state in which it thrives within the ecosystem in which it's being cared for by him and his team—he has been able to engage in the kind of social action that he has prized since his days laboring alongside people whose circumstances were largely dictated by the post-apartheid socioeconomic hangover into which they'd been born.

Once he was able to afford it, Reyneke created the Cornerstone Project, which aims to shift the ways in which the men and women who work on his land and in the winery and tasting room are treated, because for a long time, the power dynamic of post-apartheid South Africa had not morphed enough and opportunity—for stability, for growth—was still in too-short supply for too many Black and Coloured people.

The program is rooted in the idea that working the land as he does should not just be beneficial to the environment but also to the people who make that possible in the first place. So members of the staff are helped with housing and education, both for them and their children. Retirement and funeral savings are part of their pay structures, allowing them the promise of a greater sense of stability into the future than they would have had if they'd been paid as most workers historically have been. It also provides a stronger foundation for their children, which most of his employees told him was the most important facet of their work; like so many parents around the world, their hope was that their labors would provide the kind of opportunities for their children that they never had. The Cornerstone Project is a way of trying, at least among his team, to break the generational cycle of poverty that affects too many people of color even today in South Africa, more than twenty-five years after apartheid officially ended.

Programs like Cornerstone continue to grow around South Africa—some helmed by white people, some by South Africans of color, yet all of them with the ultimate goal of changing the damaging, grinding dynamic that has existed for far too long, even in the post-apartheid era. It hasn't been easy; tensions still remain, but these programs seem to be a step in the right direction. And as climate change continues to pick up speed, it will hopefully provide some sense of stability for the people most likely to be affected by it in the most damaging ways.

Wine has been produced in South Africa for more than 360 years. Its roots stretch back to the Age of Exploration, when the great ships plying the seas had to sail around the far southern tip of Africa to get from the Atlantic to the Indian Ocean or vice versa. By the time the navigators of that pre–Suez Canal era approached the bottom of the continent, it's a fair guess that they needed a good drink. The trip from Europe and England to the countries of the East, after all, was an arduous one—more than seven thousand miles from Lisbon to the Cape of Good Hope (just about halfway there!), for example, and given the fact that the ships of the time used wind and sails to power the journey, it was a long, exhausting route.

It only makes sense, then, that South Africa has been an important producer of wine since the early 1700s when Governor Simon van der Stel's Constantia estate, established earlier in 1685, led the way in building the country's reputation. Those mariners swinging around the Cape, heading from one side of the planet to the other, whether for trade or conquest—or more often than not, some toxic amalgam of both—were a perfect captive audience, a guaranteed market thirsty for everything that South Africa's vintners could produce. Exports eventually grew, and South African wine had gained an audience in England and beyond by the eighteenth century.

By the time apartheid ended, many of its grape growers and winemakers had fallen behind their international counterparts in the production of high-quality wines. This was a result of both international politics and domestic South African systems that tended to favor quantity over quality, Jim Clarke, author of *Wines of South Africa* and marketing manager for Wines of South Africa USA, told me. Despite the incredible natural advantages that the country benefits from—coastlines abutting two oceans; diverse, ancient soils; mountain ranges and valleys that allow for grape growing on well-drained hillsides of incredible complexity; a huge spectrum of climates, many of which boast the all-important diurnal swing between daytime highs and nighttime lows that allow each individual grape to both develop enough sugar during the day and maintain adequate acidity at night—South Africa had become a vinous afterthought.

Once apartheid ended, the wine world rediscovered all that it had to offer, and what they found was miraculous: from Franschhoek in the Western Cape to the more easterly stretches of the Klein Karoo, South Africa had the potential to produce world-class wine in a fantastic range of styles.

The turnaround was whiplash fast, and international investment poured into the South African wine industry. In the thirty-plus years since Nelson Mandela was released from prison, wine has grown to become an important part of the country's economy and international reputation. According to *Wines of South Africa*, "269,096 people were employed both directly and indirectly in the wine industry in 2019, including farm labourers, those involved in packaging, retailing and wine tourism. The study also concluded that of the 55 billion [Rand, or approximately $3.5 billion] contributed by the wine industry to the national GDP, about 31 billion [Rand, or around $1.9 billion] would remain in the Western Cape to the benefit of its residents."[12]

But climate change is challenging all of that progress in ways that are proving to be an existential threat to the people who can least afford it: the poor and the laborers who physically make it all possible. And all of it—the environmental, human, and economic costs of climate change—seems to be coming to a head far faster than had been predicted even ten years ago. Working more sustainably, as Reyneke told me—not just sustainably as the term refers to the actual work in the land, but sustainably from a human and economic standpoint too—is the only possible solution. "In my mind, sustainability is a three-legged chair. You have to look after nature, you have to look after people, but the hard reality of life is you also have to look after money. In fact, it's kind of unfair," he said. "You can exploit nature the longest and get away with it, people the second longest, but when you run out of money, the buck stops immediately. You have this wonky little chair on three legs, and you need to look after all three legs, but you must be very aware . . . that there are times when one will have to err on the side of financial caution."

What that financial caution means in practice, however, is constantly being redefined. For a long time, that meant thinking short term: spending money on the kind of chemicals that would effectively supercharge the land, maximizing the yields of each individual vine and making the most financially out of each growing season. It eventually became clear, however, that that was not a viable long-term strategy: it wasn't only unhealthy for the people in the vineyards working the land (Reyneke was neither the first nor the last, when he was a young man working in the fields during his grad school days, to find himself increasingly frightened by the chemicals he was being told to use), but it also wasn't sustainable financially, in many cases, for the owners of the farms and vineyards. Those herbicides and fungicides and fertilizers, of course, bumped up yields for a time, but they also often exhausted and eventually depleted the soil, which had become so denuded of natural life aside from the crops and vines themselves that it would become incapable of nourishing much growth at all without the aid of those very chemical products that depleted it in the first place. In that regard, the cycle wasn't all that different than it is for opioid-addicted people whose brains eventually function at a perceived normal level only in the presence of the chemical and fail to fire properly in its absence. In a human being and on a vineyard, that's not sustainable at all.

On his own land, this has become Johan Reyneke's passion: to break this cycle and to continue bringing life back to the patch of the planet that he tends. "We agree that global warming is real, and that our experience of it is real, and that its impacts on my business and on my lived experience" are real, he told me. "But then I have to ask myself, 'What am I doing not just to manage it, but what am I doing to facilitate or to foster or to contribute? How does my life contribute to global warming and make it worse?'" He paused. "That is a scary thought because when I did my research, I could see that there are obviously many contributing factors to global warming, but some contribute to a greater extent than others do. Agriculture," he went on, "is one of the top five. It's one of the big five contributors of global warming. That's big, man. It's not just fossil fuels and this and that. . . . It's how we grow our food and make our wine. It's how I earn my keep; it's how I spend my life on this planet. It becomes a hell of a shock to me to think and to understand that this wine industry that I love so much is part of one of the five major contributors to this very serious thing that is getting worse exponentially." With that realization, however, also came a sense of power. "The flip side of that coin is that I also have one of the biggest levers to reduce or reverse climate change. [And] with that realization has come a hell of a sense of responsibility. It's like Spider-Man says, 'With great power comes great responsibility.' You have these vast tracts of land that we're farming and growing, and if I just left nature to build topsoil, it would take five hundred years to build topsoil to the depth of a matchbox. Five hundred years! . . . And I can unplow that in half an hour through farming. It's huge. We have no idea how long it took to build this stuff that we are always just taking for granted to grow our food and make our wine in. And it's not infinite. It's *finite*, and we're running out of it fast and furiously. It's very serious."

Maybe it's his background in philosophy, or perhaps it's rooted in his experience of actually working this ancient land and seeing the way that farming as a physical and economic activity affects it as well as the people who toil on it. Regardless, Reyneke has given this situation as much thought as anyone I've ever spoken with. And for all the struggles, for all the setbacks, he seems to be making a significant impact; his work is being increasingly mirrored by others, and his position as being among the first South African winegrowers to gain biodynamic certifications from both domestic organizations and benchmark international ones like Demeter has given his voice a greater sense of gravitas when he discusses all that he's implemented.

Today, he explained, "My thinking has become less personal. It's less about me not wanting to work with chemicals, and me feeling my personal view is a softer or less aggressive approach to agriculture because that's just the way I'm wired. . . . It's actually quite serious now. Shit is real, and I'm a player in an industry where I don't really have a choice. Depending on how I get up and I live every day, I'm either making it a lot worse or a lot better. There's no middle ground. What I'm thinking is [that] agriculture can be destructive or it can be regenerative. Forget about organic or biodynamic or permaculture or all these buzzwords and catchphrases and stuff: you're either busy breaking down and destroying for profit, or you are actually doing humanity a favor by sequestering carbon on behalf of everyone out there. And that's radical, man. It's a radical thought."

How Reyneke goes about that is becoming increasingly ingrained in the life of his land and informing every decision he makes, even the least seemingly impactful. Take, for instance, the line of oak trees on his property. It's been there "for a couple of hundred years," he explained, but that doesn't mean that his decisions on how to manage and maintain them are any less important. Even when it comes to pruning them now, he's gone down the rabbit hole—he tends to do that—of what his actions and decisions would mean both now and in the future.

So he gamed it all out and now plans ahead with obsessive attention to detail. Trees are well-known consumers of carbon dioxide, which needs to be removed from the atmosphere anyway. Birds and insects find a home in them. But old trees need to be taken care of, and he knows that any number of dead branches have to be cut off to give them a chance to thrive. Pruning is necessary. "What do you do with all of that stuff? You can burn all of that stuff, and all that is going to happen is you're going to get rid of your waste but you're going to burn it and put a bunch of CO_2 into the atmosphere," he lamented. The solution, then, was to "rethink it and say, 'Maybe this is something I can use, so let's get a chipper and let's chip all the fine stuff and build that into our compost, and then we put that back into our vineyards and build humus in the process,'" which is a great solution for the smaller parts of the trees that have to be cut, but what, he questioned, of the larger logs? "There is opportunity for it," he explained. "It's called bio-char. What you do is, you make a fire but with a very low oxygen content, so you essentially don't really burn the wood; you just turn it into charcoal. And then you can go and grind that charcoal, and you can physically build it into your wall or into your structure to sequestrate the carbon forever. But what *we* do is, we pelletize it, and we infuse it with slurries, like animal slurries from cow manure, chicken manure, whatever, and then spray those in the vineyards. We are physically taking the bits and pieces of carbon and putting that back in the vineyard, but we are also giving a source of food for the vines going forward that they can draw from."

That, it seems, is the driving force behind Johan Reyneke's work: to constantly search for ways to not just increase and improve the health of his land but also to make that health more sustainable and more beneficial. He is striving to create a new kind of virtuous cycle—for both the earth and for the people working it—through the act of farming his land. It's not perfect; nothing is. But it seems to be having a far greater impact than even Reyneke could have imagined as a young man working in the fields during the day and reading his hero-philosophers at night.

NOTES

INTRODUCTION

1. Cal Fire, "Camp Fire Incident," https://www.fire.ca.gov/incidents/2018/11/8/camp-fire/.

2. "Wine & Spirits Wholesaling in the U.S.—Market Size 2003–2027," IBISWorld, modified June 7, 2021, https://www.ibisworld.com/industry-statistics/market-size/wine-spirits-wholesaling-united-states/.

3. "Alcoholic-Beverages Global Market Report 2021: COVID-19 Impacts and Forecasts to 2030—ResearchAndMarkets.com," Yahoo Finance, published August 17, 2021, https://finance.yahoo.com/news/alcoholic-beverages-global-market-report-150400574.html.

CHAPTER 1

1. Jon Bonné, "The Case of the Missing Pinot," *San Francisco Chronicle*, October 19, 2010, https://www.seattlepi.com/lifestyle/food/article/The-case-of-the-missing-Pinot-900194.php.

2. "Press," Kutch Wines, https://kutchwines.com/press-2.

3. "Press," Kutch Wines, https://kutchwines.com/press-2.

4. Julie Johnson and Mary Callahan, "Cal Fire Says Tubbs Fire Caused by Private Electrical System, Not PG&E," *The Press Democrat*, January 24,

2019, https://www.pressdemocrat.com/article/news/cal-fire-says-tubbs-fire-caused-by-private-electrical-system-not-pge/?gallery=7CDEE642-79E8-4433-B1A7-861C2DE53236.

5. "Investigation Report," California Department of Forestry and Fire Protection, Sonoma-Lake Napa Unit, May 2019, http://s1.q4cdn.com/880135780/files/doc_downloads/2019/05/TUBBS-LE80_Redacted.pdf.

6. Virginie Boone, "How California's 2017 Vintages Prevailed," *Wine Enthusiast*, August 29, 2018, https://www.winemag.com/2018/08/29/california-2017-vintages-prevailed/.

7. "Facts + Statistics," Insurance Information Institute, 2022 Update, https://www.iii.org/fact-statistic/facts-statistics-wildfires.

8. "2017 Incident Archive," Cal Fire, https://www.fire.ca.gov/incidents/2017/.

CHAPTER 2

1. Oliver Styles, "Parker and Robinson in War of Words," *Decanter.com*, April 14, 2004, https://www.decanter.com/wine-news/parker-and-robinson-in-war-of-words-102172/.

2. Encyclopaedia Britannica, "European Heat Wave of 2003," *Britannica.com*, https://www.britannica.com/event/European-heat-wave-of-2003.

3. Jelisa Castrodale, "Up to 80% of French Vineyards Have Been Damaged by Heavy Frost," *Food & Wine*, April 16, 2021, https://www.foodandwine.com/news/france-wine-vineyards-frost-damage-2021.

4. Alder Yarrow, "Imagining a Better Future for the Soils of Champagne," *Vinography*, April 27, 2015, https://www.vinography.com/2015/04/imagining_a_better_future_for.

CHAPTER 3

1. Damian Klop, "Beer As a Signifier of Social Status in Ancient Egypt with Special Emphasis on the New Kingdom Period (Ca. 1550-1069 BC): The Place of Beer in Egyptian Society Compared to Wine," SUNScholar Research Repository, Stellenbosch University Library and Information Services, 2015, https://scholar.sun.ac.za/handle/10019.1/96488.

2. Adam Montefiore, "The Wine of Israel," in *The Wine Route of Israel*, ed. Eliezer Sacks (Tel Aviv: Cordinata Publishing House, Ltd., 2015), 112.

3. Haviv Rettig Gur, "As Israel Gets Hotter and Drier, It Must Prevent Fires, Not Just Fight Them," *The Times of Israel*, September 12, 2021, https://www.timesofisrael.com/as-israel-gets-hotter-and-drier-it-must-prevent-fires-not-just-fight-them/.

4. Israel Meteorological Service, "The Rain Episode of 19 to 21 November 2020," https://ims.gov.il/en/node/1209.

5. Lahav Harkov and Tzvi Joffre, "Israel Requests International Aid Amid Largest Fire Since Carmel Blaze," *The Jerusalem Post*, August 17, 2021, https://www.jpost.com/israel-news/jerusalem-area-blaze-continues-for-second-day-residents-evacuated-676852.

6. "50 Years Ago: The Reclamation of a Man-Made Desert," *Scientific American*, February 23, 2010, https://www.scientificamerican.com/article/reclamation-of-man-made-desert/.

CHAPTER 4

1. Abby Schultz, "Rising Global Attention to California Wines," *Barron's*, August 23, 2021, https://www.barrons.com/articles/rising-global-attention-to-california-wines-01629754448.

2. Patrick Comiskey, "Who Knew England Was a Great Region for Sparkling Wine?" *Los Angeles Times*, October 25, 2017, https://www.latimes.com/food/dailydish/la-fo-english-wine-20171025-story.html.

3. Vicki Denig, "Why Do Most Champagne Bottles Lack a Vintage?" Vinepair.com, October 9, 2016, https://vinepair.com/articles/bottles-champagne-not-vintage/.

4. Mike Pomeranz, "As Harvesting Begins in Champagne, Over Half the Grapes Have Already Been Lost," Food & Wine, September 10, 2021, https://www.foodandwine.com/news/champagne-harvest-2021-weather-climate.

5. "Terroir & Appellation: The Champagne Terroir: A Dual Climate," Comité Champagne, https://www.champagne.fr/en/terroir-appellation/champagne-terroir/a-dual-climate.

6. Tony Eva, "Growing Season Heat Accumulation in Central England Since 1950," EnglishWines.info, August 18, 2015, https://englishwines.info/english-terroir/growing-season-heat-accumulation-in-central-england-since-1950/.

7. Neal Martin, "From Domesday to Now: Nyetimber," Vinous, September 2, 2021, https://vinous.com/articles/from-domesday-to-now-nyetimber-sep-2021.

CHAPTER 5

1. Meredith Daugherty, "Storm Causes Severe Damage to Central Kentucky Farms," *Bloodhorse*, July 21, 2018, https://www.bloodhorse.com/horse-racing/articles/228639/storm-causes-severe-damage-to-central-kentucky-farms.

2. Chris Schimmoeller, "Guest Columnist: Protect Historic Blanton Crutcher Farm," *The State Journal*, June 3, 2020, https://www.state-journal.com/opinion/guest-columnist-protect-historic-blanton-crutcher-farm/article_4e2f4b56-a595-11ea-bdc1-0f76d4d1e080.html.

3. Matt Strickland, "Distiller Cuts: Separating the Head, the Heart, and the Tails," Distiller.com, August 29, 2019, https://distiller.com/articles/distiller-cuts.

4. "Heads, Hearts, and Tails," StillDragon, https://stilldragon.com/blog/heads-hearts-and-tails/.

5. Clay Risen, "Rolling Out a Smaller Barrel Sooner," *The New York Times*, August 21, 2012, https://www.nytimes.com/2012/08/22/dining/whiskey-start-ups-are-rolling-out-a-smaller-barrel-sooner.html.

6. "E.H. Taylor Jr. Warehouse C Tornado Surviving," Buffalo Trace Distillery, https://www.buffalotracedistillery.com/our-brands/e-h-taylor-jr/e-h-taylor-jr-warehouse-c-tornado-surviving.html.

7. "Experimental Collection," Buffalo Trace Distillery, https://www.buffalotracedistillery.com/our-brands/experimental-collection.html.

CHAPTER 6

1. Paola Peñafiel, "Chilean Wine: 460 Years of History," Concha y Toro, https://conchaytoro.com/en/blog/chilean-wine-460-years-of-history/.

2. Tom Bruce-Gardyne, "Argentinian Malbec: A Guide to the Grape's History and Unique Style," *The Real Argentina by Bodega Argento*, April 20, 2010, https://therealargentina.com/en/argentinian-malbec-a-guide-to-the-grapes-history-and-unique-style/#:~:text=Today%20it%20is%20planted%20everywhere,miles%20(3%2C2.

3. Bruce Sanderson, "Next Wine Stop: Patagonia," *Wine Spectator*, March 31, 1998, https://www.winespectator.com/articles/next-wine-stop-patagonia-7607.

4. "Argentine & Chilean Varietals Continue Their Appeal to U.S. Wine Consumption Market," *Wine Industry Advisor*, September 1, 2020, https://

wineindustryadvisor.com/2020/09/01/argentine-chilean-varietals-appeal-us-wine.

5. "Laura Catena: Managing Director: Bodega Catena Zapata," UC Davis Wineserver, https://wineserver.ucdavis.edu/people/laura-catena#/.

CHAPTER 7

1. Andrew Weber, KUT, "Texas Winter Storm Death Toll Goes Up to 210, Including 43 Deaths in Harris County," Houston Public Media, July 14, 2021, https://www.houstonpublicmedia.org/articles/news/energy-environment/2021/07/14/403191/texas-winter-storm-death-toll-goes-up-to-210-including-43-deaths-in-harris-county/.

2. Chris Stipes, "New Report Details Impact of Winter Storm Uri on Texans," University of Houston, March 29, 2021, https://www.chicagomanualofstyle.org/tools_citationguide/citation-guide-1.html.

3. "Valentine's Week Winter Outbreak 2021: Snow, Ice, & Record Cold," National Weather Service, https://www.weather.gov/hgx/2021ValentineStorm.

4. "Winter Storm 2021 and the Lifting of COVID-19 Restrictions in Texas," University of Houston, Hobby School of Public Affairs, https://uh.edu/hobby/winter2021/.

5. "The Impact of Hurricane Harvey," University of Houston Hobby School of Public Affairs, 2020 report, https://uh.edu/hobby/harvey/.

CHAPTER 8

1. "Apartheid," history.com, October 7, 2010, https://www.history.com/topics/africa/apartheid.

2. Helen Suzman, "Key Legislation in the Formation of Apartheid," SUNY Cortland, https://www.cortland.edu/cgis/suzman/apartheid.html#:~:text=The%20Population%20Registration%20Act%20No,in%20the%20individual's%20Identity%20Number.

3. Welcome Lishivha, "The Rise of Black South African Winemakers," *Mail & Guardian*, April 23, 2019, https://mg.co.za/article/2019-04-23-the-rise-of-black-south-african-winemakers/.

4. Jim Clarke, "The Wines of South Africa," *Infinite Ideas*, July 20, 2020, Page 4.

5. Peter Weltman, "Social Justice Is Sparking Change in South Africa's Wine Industry," *SevenFifty Daily*, April 18, 2019, https://daily.sevenfifty.com/social-justice-is-sparking-change-in-south-africas-wine-industry/.

6. "Future Scenarios for the South African Wine Industry, Part 1: Impact of Climate Change," *VinIntell*, May 2012, Issue 12, http://www.sawis.co.za/info/download/VinIntell_May_2012_issue_12_part_1.pdf.

7. "*A Sand County Almanac*," Aldo Leopold Foundation, https://www.aldoleopold.org/about/aldo-leopold/sand-county-almanac/.

8. Suzanne Mustacich, "Is Copper Safe for Wine?" *Wine Spectator*, November 29, 2018, https://www.winespectator.com/articles/is-copper-safe-for-wine.

9. "Level 6 Water Restrictions Come into Effect from 1 January 2018," *Cape Town etc.*, December 5, 2017, https://www.capetownetc.com/news/level-6-water-restrictions/.

10. United States Environmental Protection Agency, "Showerheads," https://www.epa.gov/watersense/showerheads.

11. Christian Alexander, "Cape Town's 'Day Zero' Water Crisis, One Year Later," *Bloomberg CityLab*, April 12, 2019, https://www.bloomberg.com/news/articles/2019-04-12/looking-back-on-cape-town-s-drought-and-day-zero.

12. "Statistics," Wines of South Africa, https://www.wosa.co.za/The-Industry/Statistics/World-Statistics/.

BIBLIOGRAPHY

Aldo Leopold Foundation. "*A Sand County Almanac*." Accessed October 2021. https://www.aldoleopold.org/about/aldo-leopold/sand-county-almanac/.

Alexander, Christian. "Cape Town's 'Day Zero' Water Crisis, One Year Later," *Bloomberg CityLab*, April 12, 2019, https://www.bloomberg.com/news/articles/2019-04-12/looking-back-on-cape-town-s-drought-and-day-zero.

Bonné, Jon. "The Case of the Missing Pinot." *San Francisco Chronicle*, October 19, 2010. https://www.seattlepi.com/lifestyle/food/article/The-case-of-the-missing-Pinot-900194.php.

Boone, Virginie. "How California's 2017 Vintages Prevailed." *Wine Enthusiast*, August 29, 2018. https://www.winemag.com/2018/08/29/california-2017-vintages-prevailed/.

Bruce-Gardyne, Tom. "Argentinian Malbec: A Guide to the Grape's History and Unique Style," *The Real Argentina by Bodega Argento*, April 20, 2010, https://therealargentina.com/en/argentinian-malbec-a-guide-to-the-grapes-history-and-unique-style/#:~:text=Today%20it%20is%20planted%20everywhere,miles%20(3%2C2.

Buffalo Trace Distillery. "E.H. Taylor, Jr. Warehouse C Tornado Surviving." Accessed September 2021. https://www.buffalotracedistillery.com/our-brands/e-h-taylor-jr/e-h-taylor-jr-warehouse-c-tornado-surviving.html.

Buffalo Trace Distillery. "Experimental Collection." Accessed September 2021. https://www.buffalotracedistillery.com/our-brands/experimental-collection.html.

Cal Fire. "2017 Incident Archive," https://www.fire.ca.gov/incidents/2017/.

California Department of Forestry and Fire Protection, Sonoma-Lake Napa Unit. "Investigation Report," May 2019. http://s1.q4cdn.com/880135780/files/doc_downloads/2019/05/TUBBS-LE80_Redacted.pdf.

Cape Town etc. "Level 6 Water Restrictions Come into Effect from 1 January 2018," December 5, 2017, https://www.capetownetc.com/news/level-6-water-restrictions/.

Castrodale, Jelisa. "Up to 80% of French Vineyards Have Been Damaged by Heavy Frost." *Food & Wine*, April 16, 2021. https://www.foodandwine.com/news/france-wine-vineyards-frost-damage-2021.

Clarke, Jim. "The Wines of South Africa," *Infinite Ideas*, July 20, 2020, Page 4.

Comiskey, Patrick. "Who Knew England Was a Great Region for Sparkling Wine?" *Los Angeles Times*, October 25, 2017. https://www.latimes.com/food/dailydish/la-fo-english-wine-20171025-story.html.

Comité Champagne. "Terroir & Appellation: The Champagne Terroir: A Dual Climate," https://www.champagne.fr/en/terroir-appellation/champagne-terroir/a-dual-climate.

Denig, Vicki. "Why Do Most Champagne Bottles Lack a Vintage?" Vinepair.com, October 9, 2016, https://vinepair.com/articles/bottles-champagne-not-vintage/.

Eva, Tony. "Growing Season Heat Accumulation in Central England Since 1950." EnglishWines.info, August 18, 2015. https://englishwines.info/english-terroir/growing-season-heat-accumulation-in-central-england-since-1950/.

Harkov, Lahav, and Tzvi Joffre. "Israel Requests International Aid Amid Largest Fire Since Carmel Blaze." *The Jerusalem Post*, August 17, 2021. https://www.jpost.com/israel-news/jerusalem-area-blaze-continues-for-second-day-residents-evacuated-676852.

History.com, "Apartheid," October 7, 2010, https://www.history.com/topics/africa/apartheid.

IBISWorld. "Wine & Spirits Wholesaling in the U.S.—Market Size 2003–2027." June 7, 2021. https://www.ibisworld.com/industry-statistics/market-size/wine-spirits-wholesaling-united-states/.

Insurance Information Institute. "Facts + Statistics: Wildfires." 2022 Update. https://www.iii.org/fact-statistic/facts-statistics-wildfires.

Israel Meteorological Service, "The Rain Episode of 19 to 21 November 2020," https://ims.gov.il/en/node/1209.

Johnson, Julie, and Mary Callahan. "Cal Fire Says Tubbs Fire Caused by Private Electrical System, Not PG&E." *The Press Democrat*, January 24, 2019. https://www.pressdemocrat.com/article/news/cal-fire-says-tubbs-fire-caused-by-private-electrical-system-not-pge/?gallery=7CDEE642-79E8-4433-B1A7-861C2DE53236.

Klop, Damian. "Beer As a Signifier of Social Status in Ancient Egypt with Special Emphasis on the New Kingdom Period (Ca. 1550-1069 BC): The Place of Beer in Egyptian Society Compared to Wine," SUNScholar Research Repository, Stellenbosch University Library and Information Services, 2015, https://scholar.sun.ac.za/handle/10019.1/96488.

Kutch Wines. "Press." Accessed July 2021. https://kutchwines.com/press-2.

Lishivha, Welcome. "The Rise of Black South African Winemakers." *Mail & Guardian*, April 23, 2019. https://mg.co.za/article/2019-04-23-the-rise-of-black-south-african-winemakers/.

Martin, Neal. "From Domesday to Now: Nyetimber," Vinous, September 2, 2021, https://vinous.com/articles/from-domesday-to-now-nyetimber-sep-2021.

Montefiore, Adam. "The Wine of Israel." In *The Wine Route of Israel*, edited by Eliezer Sacks, 12. Tel Aviv: Cordinata Publishing House, Ltd., 2015.

Mustacich, Suzanne. "Is Copper Safe for Wine?" *Wine Spectator*, November 29, 2018. https://www.winespectator.com/articles/is-copper-safe-for-wine.

National Weather Service. "Valentine's Week Winter Outbreak 2021: Snow, Ice, & Record Cold." Accessed October 2021. https://www.weather.gov/hgx/2021ValentineStorm.

Peñafiel, Paola. "Chilean Wine: 460 Years of History." Concha y Toro. https://conchaytoro.com/en/blog/chilean-wine-460-years-of-history/.

Pomeranz, Mike. "As Harvesting Begins in Champagne, Over Half the Grapes Have Already Been Lost," *Food & Wine*, September 10, 2021, https://www.foodandwine.com/news/champagne-harvest-2021-weather-climate.

Rettig Gur, Haviv. "As Israel Gets Hotter and Drier, It Must Prevent Fires, Not Just Fight Them," *The Times of Israel*, September 12, 2021, https://www.timesofisrael.com/as-israel-gets-hotter-and-drier-it-must-prevent-fires-not-just-fight-them/.

Risen, Clay. "Rolling Out a Smaller Barrel Sooner," *The New York Times*, August 21, 2012, https://www.nytimes.com/2012/08/22/dining/whiskey-start-ups-are-rolling-out-a-smaller-barrel-sooner.html.

Sanderson, Bruce. "Next Wine Stop: Patagonia." *Wine Spectator*, March 31, 1998. https://www.winespectator.com/articles/next-wine-stop-patagonia-7607.

Schultz, Abby. "Rising Global Attention to California Wines." *Barron's*, August 23, 2021. https://www.barrons.com/articles/rising-global-attention-to-california-wines-01629754448.

Scientific American. "50 Years Ago: The Reclamation of a Man-Made Desert." February 23, 2010. https://www.scientificamerican.com/article/reclamation-of-man-made-desert/.

Schimmoeller, Chris. "Guest Columnist: Protect Historic Blanton Crutcher Farm," *The State Journal*, June 3, 2020, https://www.state-journal.com/opinion/guest-columnist-protect-historic-blanton-crutcher-farm/article_4e2f4b56-a595-11ea-bdc1-0f76d4d1e080.html.

StillDragon.com. "Heads, Hearts, and Tails," https://stilldragon.com/blog/heads-hearts-and-tails/.

Stipes, Chris. "New Report Details Impact of Winter Storm Uri on Texans." University of Houston, March 29, 2021. https://www.chicagomanualofstyle.org/tools_citationguide/citation-guide-1.html.

Strickland, Matt. "Distiller Cuts: Separating the Head, the Heart, and the Tails," Distiller.com, August 29, 2019, https://distiller.com/articles/distiller-cuts.

Suzman, Helen. "Key Legislation in the Formation of Apartheid," SUNY Cortland, https://www.cortland.edu/cgis/suzman/apartheid.html#:~:text=The%20Population%20Registration%20Act%20No,in%20the%20individual's%20Identity%20Number.

United States Environmental Protection Agency, "Showerheads," https://www.epa.gov/watersense/showerheads.

University of California Davis Wineserver, "Laura Catena: Managing Director: Bodega Catena Zapata," https://wineserver.ucdavis.edu/people/laura-catena#/

University of Houston Hobby School of Public Affairs "The Impact of Hurricane Harvey," 2020 report, https://uh.edu/hobby/harvey/.

University of Houston, Hobby School of Public Affairs. "Winter Storm 2021 and the Lifting of COVID-19 Restrictions in Texas." Accessed October 2021. https://uh.edu/hobby/winter2021/.

VinIntell. "Future Scenarios for the South African Wine Industry, Part 1: Impact of Climate Change." May 2012, Issue 12. http://www.sawis.co.za/info/download/VinIntell_May_2012_issue_12_part_1.pdf.

Weber, Andrew. KUT, "Texas Winter Storm Death Toll Goes Up to 210, Including 43 Deaths in Harris County," Houston Public Media,

July 14, 2021, https://www.houstonpublicmedia.org/articles/news/energy-environment/2021/07/14/403191/texas-winter-storm-death-toll-goes-up-to-210-including-43-deaths-in-harris-county/.

Weltman, Peter. "Social Justice Is Sparking Change in South Africa's Wine Industry." *SevenFifty Daily*, April 18, 2019. https://daily.sevenfifty.com/social-justice-is-sparking-change-in-south-africas-wine-industry/.

Wine Industry Advisor. "Argentine & Chilean Varietals Continue Their Appeal to U.S. Wine Consumption Market." September 1, 2020. https://wineindustryadvisor.com/2020/09/01/argentine-chilean-varietals-appeal-us-wine.

Wines of South Africa. "Statistics." The Industry. Accessed November 2021. https://www.wosa.co.za/The-Industry/Statistics/World-Statistics/.

Yahoo Finance. "Alcoholic-Beverages Global Market Report 2021: COVID-19 Impacts and Forecasts to 2030—ResearchAndMarkets.com." August 17, 2021. https://finance.yahoo.com/news/alcoholic-beverages-global-market-report-150400574.html.

Yarrow, Alder. "Imagining a Better Future for the Soils of Champagne." Vinography, April 27, 2015. https://www.vinography.com/2015/04/imagining_a_better_future_for.

INDEX

Aaron, Stu, 101–4
ABV, of whiskey, 93
acids in wine: in California, 9; in Israel, 52
ACTivation, 102, 103, 104
Adama Wines, 157
agave, fertilizer from, xviii
aging: of gin, 96–97; of whiskey, 93, 96, 99–101, 108–9; of wines, 22–23, 29
Akerman, Michal, 43–55
alcohol, 24; in Bordeaux, 73; in bourbon, 93; in Merlot, 36; rising levels of, 6; in whiskey, 93–94
aldehydes, in whiskey distillation, 92–93
Anakota, 20
Andes Mountains, 112–13; planting in, 124; snowmelt from, 120–22
Angel's Envy, 93–94
Anglo-Paris Basin, 66
apartheid, in South Africa, 156, 161–62, 174–75
Argentina. *See* Patagonia
Arinarnoa, 28
Arvin, Will, 90
Aslina, 157
Australia: Barossa Valley of, 21, 36; Israel and, 60; Shiraz of, 85
Australian Wine Research Institute, 127

Bacardi, xviii
Bacchus, 76, 79
Balfour Winery, 70, 85
Banke, Barbara, 20, 23
Barkan Winery, 60
Barossa Valley, of Australia, 21, 36
barrels. *See* oak barrels
Barzi, Guillermo, 114–15
Bayede!, 157
Bellevue, 19

Bell Mountain American Viticultural Area, 145
Bending Branch Winery, 149–59
Bergamot, 97
Bespoken Spirits, xvi, 101–4
bio-char, 177
biodiversity, in Israel, 44, 48
biodynamics: in Bordeaux, 34; in Israel, 45; in South Africa, 163–64
biofertilizer, in Patagonia, 131
Biyela, Ntsiki, 156–57
the black wine *(le vin noir)*, 113
Blanc, 27
Blanton's, 94
Bloom Energy, 102
Bloublommetjieskloff, 165
Bodega Catena Zapata, 124–27, 131
Bodega Chacra, 117
Bodega y Cavas de Weinert, 117
Bollinger, 77, 80
Bonné, Jon, 8
Boone, Virginie, 12
Booth's, 97
Bordeaux: alcohol in, 73; biodynamics in, 34; Cabernet Franc of, 28, 146; Cabernet Sauvignon of, 45, 146; California and, 29; challenging weather of, 26–27, 29; climate change in, 17–37; cold snaps in, 37; England and, 73; fermentation in, 32; First Growths of, 64, 124; grand châteaux of, 19; hail in, 30–31, 37; herbicides in, 32–33; Israel and, 60; The Judgment of Paris and, 63; Malbec of, 28; Merlot of, 22, 45, 146; new permitted varieties in, 148–49; old vines in, 23, 25; permitted seed grapes in, 28; Port and, 29; regulated appellations of, 26; sustainable farming in, 34
Bordeaux AOC, 28
Bordeaux Supérieur, 27, 28
botanicals, in gin, 98
The Botanist, 97
Botrytis cinerea fungus, 27
Bottle Shock, 63
bourbon, 92; aging of, 96; alcohol in, 93; of Castle & Key, 96; in oak barrels, 108
brandy, 97
Branson, Richard, 73
Brazil, 113
Bride Valley, 68, 69–70, 85
Browne, Michael, 5
Bruichladdich, 97
Brut NV, 77, 80
Buffalo Trace, 94, 108–9
Burgundy: California and, 6–12; Chardonnay of, 2, 45, 146; The Judgment of Paris and, 63; Pinot Noir of, 2, 4, 45, 122, 146; Romanée-Conti in, 106; shortages of, xv; stem inclusion in, 8
buttonhole, of clay, 27

Cabernet: of Harlan Estate, 5; of Margaret River, 85
Cabernet Franc: of Bordeaux, 28, 146; of Château Lassègue, 22, 25–26, 31, 35, 36; genetic mutation of, 78; of Le Désir, 20; of the Right Bank, 26; of Saint-Émilion, 26, 27
Cabernet Sauvignon, 14; of Bordeaux, 28, 45, 146; of California, 147; of Château

Lassègue, 22, 26, 31–32, 35, 36; cryo-maceration of, 151; development of, 31–32; genetic mutation of, 78; of Humberto Canale, 115–16; of Israel, 55; The Judgment of Paris and, 63; of La Joie, 20; of the Left Bank, 27; of Patagonia, 114, 115–16, 124; of Stag's Leap Wine Cellars S.L.V., 63; of Texas Hill Country, 147, 151; from *Vitis vinifera,* 112
Les Cadrans, 22, 26
California: Bordeaux and, 29; Burgundy and, 6–12; Cabernet Sauvignon of, 147; climate change in, 1–15; cult wines of, 64; diurnal shift in, 9; Domaine Carneros of, 84; droughts in, 13; fermentation in, 9, 12; global wine trade share of, 64; heatwaves in, 13; The Judgment of Paris and, 64; oak barrels in, 24; pick date in, 7; Pinot Noir of, 5, 7, 13–14; rising alcohol levels in, 24; smokiness of wines in, 8; Sonoma Valley in, 1–15; stem inclusion in, 8; tendril clipping in, 9; terroir in, 13–14, 19; wildfires in, xiv, 8, 11, 12–13, 49, 76. *See also specific areas and topics*
camels, in Israel, 58
Camp Fire, xiv, 13
Canale, Humberto, 115
carbon sequestration: in Patagonia, 131; in South Africa, 177–78
Carignan, 56
Carlos V (King), 112
Carmel Winery, 58
Carménère, 28
Castets, 28
Castle & Key: bourbon of, 96; floods of, 87–90, 110; head blender at, 90
Catena, Laura, 124–25, 126–27
Catena Institute of Wine, 126
Cava, 72
cellar master *(chef de cave),* 67, 77
Central Valley, of Patagonia, 65
Champagne, 45; compost from trash in, 33–34; England and, 66–68, 77–80, 83, 86; great houses of, 19; herbicides in, 32–33; NM on, 77; NV, 67; RM on, 77; smallest harvest of, xv
Champagne Salon, 4
Chardonnay: of Bodega Chacra, 117; of Bodega y Cavas de Weinert, 117; of Bride Valley, 69–70; of Burgundy, 2, 45, 146; Champagne and, 67; of Coche-Dury Meursault, 2, 4; of Digby Fine English, 84; of England, 68, 74, 76, 79, 83; The Judgment of Paris and, 63; of Kutch Wines, 10; from Napa Valley, 63; nature of, 2–3; of Nyetimber, 73; of Patagonia, 124; of Texas Hill Country, 147; from *Vitis vinifera,* 112
Château Angelus, 26
Château Cheval Blanc, 26
Château Haut-Brion, 27, 63
Château Lafite, 58
Château Lafite Rothschild, 26
Château Lassègue, 17–37; aging of wines at, 22–23; Cabernet Franc of, 22, 25–26, 31, 35, 36; Cabernet Sauvignon of, 22, 26, 31–32, 35, 36; consultants at, 21,

25; *grand vin* of, 22, 35; harvest at, 35; limestone of, 22; Merlot of, 18, 31–32, 35, 36; older vines at, 25–26; sundials of, 22
Château Latour, 26
Château Léoville-Las Cases, 63
Château Margaux, 26
Château Montrose, 63
Château Pétrus, 26
Châteaux Mouton Rothschild, 63
chef de cave (cellar master), 67, 77
Chile. *See* Patagonia
Chile, Miguel Torres, 118
citrus, in gin, 97
Clarke, Jim, 157, 173, 174
clay: buttonhole of, 27; in England, 66; in Texas Hill Country, 139
climate change: in Bordeaux, 17–37; in California, 1–15; in England, 63–86; in France, 69; in Israel, 39–62; in Kentucky, 87–110; in Patagonia, 111–32; in South Africa, 155–78; speed of, xiv–xv; in Texas Hill Country, 131–43; whiskey and, 87–110
clonal (massal) selection: in England, 70–71; in Patagonia, 126
Clough, Trevor, 76–77, 79, 80–84, 85–86
Coche-Dury Meursault, 2, 4
cold snaps: in Bordeaux, 23, 37; in Texas Hill Country, 133–34, 135–37, 141–1423, 183n1
the Comité Champagne, 67–68, 69
Concha y Toro, 112, 114
Connors, Brett, 88, 90–91
consultants, 21, 25
Cooper, John, 107
copper sulfate and lime, 165–66
Cornerstone Project, in South Africa, 164, 171–72
Corton-Charlemagne Grand Cru, 2, 3
Côt, 113
COVID, xv, xviii; in South Africa, 159
craft distillers, of whiskey, 94, 96
cryo-maceration, 151–52
cult wines, 64

Dad's Hat Rye, 107
Dalla Valle Maya, 64
Day Zero, in South Africa, 169
Demeter certification, 160
dendrometers, 43
Denmark, xv, 114
Le Désir, 20
Digby Fine English, 76, 80–86
distillation: at Bespoken Spirits, 101; of whiskey, 92–93
diurnal shift: in California, 9; in Israel, 42; in Patagonia, 123–24; in South Africa, 174
Domaine Carneros, 84
Domaine de la Romanée-Conti (DRC), 2–3
Domaine Jean François Coche-Dury, 2
Domaine Leflaive Puligny-Montrachet Les Purcelles, 63–64
Domesday Book, 73
Don Melchor, 114
Dornfelder, 79
dosage, 67
DRC. *See* Domaine de la Romanée-Conti
drip irrigation: in Israel, 43; in Patagonia, 121, 126–27
drones, 43

droughts: in California, 13; in Texas Hill Country, 153
Duchman Family Winery, 145
Duckhorn Wine Co., 5
Duero, Ribera del, 139
DWinter Storm Uri, in Texas Hill Country, 141–43, 183n1

Eagle Rare, 94
EarthOptics, xvi
earthquakes, in Patagonia, 76
Eaves, Marianne, 90
EFSA. *See* European Food Safety Authority
Egypt, 39, 180n1
E. H. Taylor, Jr.Warehouse C Tornado Surviving Bourbon, 108
England: Bordeaux and, 73; Champagne and, 66–68, 77–80, 83, 86; Chardonnay of, 68, 74, 76, 79, 83; climate change in, 63–86; clonal selection in, 70–71; GDD in, 71–72; heatwaves in, 72–73; investment in, 71; *négociant* in, 77–78, 80; Pinot Meunier of, 74, 79; Pinot Noir of, 68, 74, 76, 79, 83; rain in, 76; ripening in, 77; Romans in, 66; site selection in, 70–71; sparkling wines of, 65–86; terroir of, 66. *See also specific areas and topics*
EnglishWines.info, 71
Entre-Deux-Mers, 27
Epernay, 66
ethyl acetate, in whiskey distillation, 92–93
European Food Safety Authority (EFSA), 165
Eva, Tony, 71
Experimental Collection, 108–9

FAIR'N GREEN, 60
FAIR Spirits, xviii
Falstaff Vineyard, 10, 13
fermentation: at Bespoken Spirits, 101; in Bordeaux, 32; in California, 9, 12; cryo-maceration and, 151–52; in Israel, 53; sugar after, 123; of whiskey, 92, 106–7
fertilizer: from agave, xviii; from compost, 33; in Israel, 46; in Patagonia, 131; from snails, 166; in South Africa, 175; whiskey and, 110
First Growth Bordeaux, 64, 124
floods: of Castle & Key, 87–90, 110; in Israel, 49, 52; in Texas Hill Country, 153
Food & Wine, 30, 67
Four Roses, 100
France: climate change in, 69; heatwaves in, 23, 29–30; Israel and, 61; The Judgment of Paris and, 64; Malbec of, 113; Southern Rhône in, 124. *See also specific areas and topics*
Franciacorta, 72
Frey Ranch, 109
Friesen Vineyard, 140
fungicides, in South Africa, 162, 165, 175

Galil Mountain, 60
Galloni, Antonio, 10, 123
Ganley-Roper, Nora, 105–6, 109, 110
Garage de Papa, 60
Gascony, 20
Gastaud-Gallagher, Patricia, 68
GDD. *See* growing degree days
genetic mutation, 78–79

Germany, 79
gin, 96–98
glycerin, 7
Golan Heights, 42, 57
Golan Heights Winery, 59, 60
Good Guy, 138, 140
Göpfert, Barbara Wolff, 127–31
Grand Cru, 126
grand vin, of Château Lassègue, 22, 35
gravel, of the Left Bank, 26–27
Graves, 27
Grenache, 14; of Israel, 55; of McLaren Vale, 85; of Texas Hill Country, 133
growing degree days (GDD), 71–72
"Growing Season Heat Accumulation in Central England Since 1950," 71

hail: in Bordeaux, 29, 30–31, 37; in Israel, 55; in Saint-Émilion, 17–18, 22, 24; in Texas Hill Country, 153
hamotzi, 40
Hampshire, 66
Harlan Estate, 5, 64
head, in whiskey distillation, 92–93
head blender: at Castle & Key, 90; at Digby Fine English, 84
hearts, in whiskey distillation, 92–93
heatwaves, xiv; in California, 13; in England, 72–73; in France, 23, 29–30; in Israel, 54; in Texas Hill Country, 153
heirloom grains, for whiskey, 106–7
Helfensteiner, 79
herbicides: in Bordeaux, 32–33; in Champagne, 32–33; in Israel, 55; in South Africa, 162, 165, 175; whiskey and, 109–10
Heroldrebe, 79
Hidden Vineyard, 77, 80
House of BNG, 157
House of Mandela, 157
Houston Livestock Show and Rodeo Wine Competition, 140, 151
Humberto Canale, 114–16
humidity: in Bordeaux, 27; in Israel, 57; whiskey aging and, 99
Hurricane Harvey, 152–53

Imvula Wine, 157
Indian Springs Calistoga, xiii
insects, xiv
investment: in England, 71; in South Africa, 174; in water management, 125–26
irrigation: in Israel, 51; in Patagonia, 118, 120–22, 126–27
Islay Barley, 107
Israel, 7; biodiversity in, 44, 48; Cabernet Sauvignon of, 55; camels in, 58; Carignan in, 56; climate change in, 39–62; damaged land of, 50–51; dendrometers in, 43; diurnal shift in, 42; drip irrigation in, 43; drones in, 43; fertilizer in, 46; floods in, 49, 52; Grenache in, 55; hail in, 55; heatwaves in, 54; herbicides in, 55; higher elevated sites in, 57–58; improved wine quality in, 60–61; irrigation in, 42, 51; kosher law in, 40–41, 46; landscape and climate differences in, 42; Malbec of, 55; Marselan of, 55, 56; Master

of Wine of, 60; old vines in, 46–48; organic farming in, 45; pesticides in, 55; Petite Sirah of, 55; phenolic ripeness in, 52–54; Pinot Noir of, 55; planting and replanting in, 46; rain in, 49; reduced grape production in, 54; sommeliers and, 60; sustainable farming in, 45; technology of, 42–43, 62; in tough neighborhood, 50; wildfires in, 49, 52, 55; wine education in, 59–60
Italy, 72

Jack Daniel's, 94
Jackson, Jess, 20, 23
Jackson Family Wines, xvi–xvii
Janousek, Martin, 101–4
Japanese whiskey, 92
Jeter, Derek, 101
Jews. *See* Israel
Jim Beam Black, 94
La Joie, 20
Joseph Drouhin Beaune Clos des Mouches, 63
Judgment of Paris (Taber), 63
The Judgment of Paris, 63–64, 68
Justin's House of Bourbon, 95, 105

kashrut, 40, 46
Kent, 66, 70, 77, 80, 84
Kentucky straight bourbon whiskey, 93
Klein Karoo, 174
Kleynhan, Vivian, 157
kosher law, 40–41, 46
Krug, 80
Kuhlken, David, 144
Kuhlken, Julie, 144–52, 154
Kutch, Jamie, 1–15
Kutch Wines, 4, 12; Chardonnay of, 10; Pinot Noir of, 9–10

the Left Bank: Cabernet Sauvignon of the, 27; gravel of the, 26–27; limestone of, 26–27; *Wine Spectator* on, 29
Leopold, Aldo, 160–61, 171
Lewinsohn, Ido, 60
Lightning Complex fires, 8
limestone: of Château Lassègue, 22; of the Left Bank, 26–27; of Saint-Émilion, 27; of Texas Hill Country, 139
Lithathi Wines, 157
LODI RULES Sustainable Winegrowing Program, 60
Loire Valley, 20
Lost Lantern, 105–6, 110
Lourie, Amichai, 39, 46
Lujan de Cujo, 117
LVMH, 107–8

Maker's Mark 46, 94
Malbec: of Bordeaux, 28; of France, 113; of Israel, 55; of Patagonia, 85, 125
Malherbe, Jeanne, 165
Mandela, Nelson, 156, 171, 174
Manigold, Edward, 139
Manigold, Madeleine, 139
Manischewitz, 41
Margaret River, 85
Marselan, 28, 55, 56
Martin, Neal, 73
mash bill, for whiskey, 91–92, 93

massal (clonal) selection:
in England, 70–71; in Patagonia, 126
Master of Wine, of Israel, 60
McDougall Ranch, 10, 13
McLaren Vale, 85
mealybug, in South Africa, 166
Médoc, 26
Mendoza, 65, 113, 117, 120–24
Merlot: of Bodega y Cavas de Weinert, 117; of Bordeaux, 22, 28, 45, 146; of Château Lassègue, 18, 31–32, 35, 36; of Château Pétrus, 26; cryo-maceration of, 151; development of, 31–32; of La Muse, 20; of Pomerol, 27; of the Right Bank, 26; of Texas Hill Country, 147, 151
methanol, in whiskey distillation, 92–93
MGP. *See* Midwest Grain Products
M'Hudi Wines, 157
Michelin, 123
microclimates: of Israel, 52; of Patagonia, 119
Midwest Grain Products (MGP), 95, 96
Mihalich, Herman, 107
Miles', 97
Moët & Chandon, 77
Moët Hennessy USA, xvi–xvii
Mondavi, Robert, 125
Montefiore, Adam, 7, 40, 56–62
Moss, Sandy, 73
Moss, Stuart, 73
Müller-Thurgau, 79
Murry, Wes, 90
Muscadelle, 27
La Muse, 20
Napa Valley, xiii–xiv; Chardonnay from, 63; cult wines of, 64; Israel and, 60
Negev, 42, 57, 58
négociant, 77–78, 80
négociant manipulant (NM), on Champagne, 77
Newsom Vineyards, 153
NM. *See négociant manipulant*
Noah, 40
Noble Rot, 64–65
non-GMO, whiskeys, 95–96
non-vintage (NV): Champagne, 67; sparkling wines, 82
Norway, 114
NV. *See* non-vintage
Nyetimber, 73–74

oak barrels: at Bespoken Spirits, 103–4; bourbon in, 108; in California, 24; whiskey in, 93–94, 99–101
old vines: in Bordeaux, 23, 25; in Israel, 46–48; in Patagonia, 126
100-point scoring system, 123
organic farming: in Israel, 45; in South Africa, 163–68; whiskey and, 110

Pacific Gas & Electric (PG&E), xiii–xiv
País, 113
Pappy Van Winkle, 94, 108
Parker, Robert, 6, 10, 20, 122
Passover, 40, 51
Patagonia (Argentina and Chile), 14, 65; Andes Mountains and, 112–13, 120–22, 124; biofertilizer in, 131; Cabernet Sauvignon of, 114, 115–16, 124; carbon

sequestration in, 131; Central Valley of, 65; Chardonnay of, 124; climate change in, 111–32; diurnal shift in, 123–24; drip irrigation in, 121; earthquakes in, 76; exports from, 122; fertilizer in, 131; irrigation in, 118, 120–22, 126–27; Israel and, 54; Malbec of, 85, 125; massal selection in, 126; microclimates of, 119; old vines in, 126; Petit Verdot in, 116; Pinot Noir of, 117, 124; rain in, 117–18; renewable energy in, 131; sustainable farming in, 127–31; terroir of, 119; winds of, 117
Patrón, xviii
Paulus, Caroline, 95, 96, 105
PD. *See* Pierce's disease
Pedernales Cellars, 144–45
Pessac-Léognan, 27
pesticides: in Bordeaux, 32; in Israel, 55; in South Africa, 162, 165; whiskey and, 109–10
Petite Sirah, 55
Petit Meslier, 67
Petit Verdot: of Bordeaux, 28; of Patagonia, 116
Pétrus, 27
PG&E. *See* Pacific Gas & Electric
phenolic ripeness: in Israel, 52–54; in Texas Hill Country, 150
photosynthesis, 24
Pierce's disease (PD), 140, 147
Piero Incisa dellla Rocchetta, 117
Pinot Blanc, 78
Pinot Gris, 78
Pinot Meunier, 67, 68; of Digby Fine English, 84; of England, 74, 79; of Nyetimber, 73
Pinot Noir: of Bodega Chacra, 117; of Bodega y Cavas de Weinert, 117; of Bride Valley, 70; of Burgundy, 2, 4, 45, 122, 146; of California, 5, 7, 13–14; Champagne and, 67; of Digby Fine English, 84; of DRC, 2–3; of England, 68, 74, 76, 79, 83; of Falstaff Vineyard, 13; genetic mutation of, 78; of Israel, 55; of Kutch Wines, 9–10; of McDougall Ranch, 10, 13; nature of, 2–3; of Nyetimber, 73; of Patagonia, 117, 124; of Romanée-Conti, 106; of Texas Hill Country, 148; from *Vitis vinifera,* 112
plant viruses, xiv; PD, 140, 147
Polonski, Adam, 105–6, 109
polyphenols, 150–51
Pomerol, 22, 26, 27
Population Registration Act of 1950, in South Africa, 156
Port, 29
Portugal, 54
Prewitt, Brian, 94–96, 97–98
pruning: of hail damage, 21–22; in South Africa, 177
Puento Alto, 114
Puligny-Montrachet Les Enseignères, 2
Purisima Creek Redwoods Open Space Preserve, 1–2

quinoa, for vodka, xviii

rain, xiv; in Bordeaux, 29; in England, 76; in Israel, 49; in Patagonia, 117–18; in South Africa, 170

Ramonet-Prudhon Bâtard-Montrachet, 64
Rattet, Michel-Henri, 30
récoltant manipulant (RM), 77
Reilly, Dave, 145, 154
renewable energy, in Patagonia, 131
Reserve Brut, 83
Revivallist Gin, 97
Reyneke, Johan, 159–78
Richebourg vineyard, 4
Ridgeview, 74
Riesling, 79, 117
the Right Bank, 26
RM. *See récoltant manipulant*
Robinson, Jancis, 10
Romanée-Conti, 106
Romans, in England, 66
"Rooted for Good," xviii
Rothschild, Edmond de, 58, 61
Roulot Meursault Charmes, 64
Roundup, 45
rye whiskey, 92, 107

sabbatical year *(shmitta),* 46, 54
Saint-Émilion, 17–37; Cabernet Franc of, 26, 27; hail in, 17–24; limestone of, 27
A Sand County Almanac (Leopold), 160–61
Sauternes, 27
Sauvignon Blanc: of Entre-Deux-Mers, 27; genetic mutation of, 78
Sazerac, 94, 95–96, 97–98
Scotch, 92, 107
Scotland, 97
Screaming Eagle, 64
Seillan, Hélène, 20
Seillan, Monique, 23
Seillan, Nicolas, 17–37
Seillan, Pierre, 18, 25, 31, 34–37; as vigneron, 19, 20, 28
Sémillon: of Entre-Deux-Mers, 27; of Texas Hill Country, 133
Ses' Fikile Wines, 157
SevenFifty Daily, 157–58
Seven Sisters, 157
Shiloh Winery, 39, 46
Shiraz, 85
shmitta (sabbatical year), 46, 54
Silvaner, 79
Sine Qua Non, 64
Smit, Stefan, 157–58
smokiness of wines, 8
snails, in South Africa, 166
snowmelt, from Andes Mountains, 120–22
Society for the Protection of Nature in Israel (SPNI), 43–44, 45, 49–50
sommeliers: English sparkling wines and, 65; Israel and, 60; Kutch Wines and, 10; South Africa and, 156
Sonoma Valley, 3, 7, 10, 14, 20, 26, 36
South Africa: apartheid in, 156, 161–62, 174–75; biodynamics in, 163–64; carbon sequestration in, 177–78; climate change in, 155–78; Cornerstone Project in, 164, 171–72; Day Zero in, 169; diurnal shift in, 174; fertilizer in, 175; fungicides in, 162, 165, 175; herbicides in, 162, 165; investment in, 174; organic farming in, 163–68; pesticides in, 162, 165; Population Registration Act of 1950 in, 156; pruning in, 177; rain in, 170; sugar in, 174;

sustainable farming in, 174–75; water management in, 168–69
South African Fruit and Wine Initiative, 159
South Downs, 73, 74
Southern Rhône, 124
Spain: Cava in, 72; Israel and, 54; Tempranillo of, 139
sparkling wines, 7; of England, 65–86; NV, 82
Spicewood Vineyards, 131–43
SPNI. *See* Society for the Protection of Nature in Israel
Spurrier, Bella, 68–69
Stagg, 94, 108
Stag's Leap Wine Cellars S.L.V., 63
Stel, Simon van der, 173
Stellekaya winery, 156–57
Stellenbosch Vineyards, 156, 157–58
stem inclusion, 8
Stern, Howard, 45
Strickland, Matt, 93
Suckling, James, 123
sugar: in California, 9; after fermentation, 123; in Israel, 52–54; from photosynthesis, 24; in South Africa, 174; in Texas Hill Country, 149–50
Sugrue, Dermot, 72–77, 82
Summer Equinox, 97
sundials, of Château Lassègue, 22
Super-Tuscan Sassicaia, 117
Surrey, 66
Sussex, 66, 76
sustainable farming: in Bordeaux, 34; in Israel, 45; in Patagonia, 127–31; in South Africa, 174–75
Suzman, Helen, 156
Syrah, 14; of Texas Hill Country, 133
Taber, George, 63
Tabor Winery, 43–44, 48–49, 52, 60
tails, in whiskey distillation, 92–93
Taittinger, 77, 84
tannins, 22–23, 29
Taylor, Edmund Haynes, Jr., 90, 108
Tempranillo: of Friesen Vineyard, 140; of Spain, 139; of Texas Hill Country, 133, 139, 140
tendril clipping, 9
Tennessee whiskey, 92
Tenuta di Arceno, 20
Tenuta San Guida, 117
Tequila Cazadores, xviii
Teroldego grapes, 148
terroir: of California, 13–14, 19; of England, 66; of Israel, 52; of Patagonia, 119
Texas Hill Country, xv; Cabernet Sauvignon of, 147, 151; Chardonnay of, 147; clay in, 139; climate change in, 131–43; cold snaps in, 133–34, 135–37, 141–43, 183n1; cryo-maceration in, 151; droughts in, 153; floods in, 153; Grenache of, 133; hail in, 153; heatwaves in, 153; Hurricane Harvey in, 152–53; limestone in, 139; Merlot of, 147, 151; phenolic ripeness in, 150; Pinot Noir of, 148; Sémillon of, 133; sugar in, 149–50; Syrah of, 133; Tempranillo of, 133, 139, 140; warming event in, 143; weather variation in, 134–35; Winter Storm Uri in, 133–37, 141–43, 183n1
Thandi Wines, 157
Thoreau, Henry David, 160–61, 171

Tinkerman's, 97
Touriga Nacional, 28
trichoderma, 166
Tubbs Fire, 11, 12–13
Tzora Vineyards, 60

Uco Valley, 119–20, 124
Uiktzicht, 171
University of California, Davis, 59, 125; Young, R., and, 151
Uruguay, 113–14

Valdivia, Pedro de, 112
vanilla, in gin, 98
Vérité, 20
Veuve Clicquot, 77, 80
Vidirí, Juan Martin, 114–16, 117
vigneron, Seillan, P., as, 19, 20, 28
Viña San Pedro - Tarapacáa Wine Group (VSPT Wine Group), 127–31
Vinepair.com, 67
Vinexpo 2020, xvi–xvii
VinIntell, 158–59
le vin noir (the black wine), 113
Vinography, 32–33
Vinous, 10, 73, 123
Vitis vinifera, 112, 113
vodka: quinoa for, xviii; *versus* whiskey, 97
VSPT Wine Group. *See* Viña San Pedro - Tarapacáa Wine Group

Walden Pond (Thoreau), 160–61
warming event, in Texas Hill Country, 143
water management: in Patagonia, 125–26; in South Africa, 168–69
Weber, Hubert, 117
Weller, 94, 108
Welmoed winery, 156
wheated bourbons, 92
whey, xviii
Wheyard Spirit, xviii
whiskey: ABV of, 93; aging of, 93, 96, 99–101, 108–9; alcohol in, 93–94; of Castle & Key, 87–110; climate change and, 87–110; craft distillers of, 94, 96; distillation of, 92–93; fermentation of, 92, 106–7; fertilizer and, 110; grains in, 94–95; heirloom grains for, 106–7; herbicides and, 109–10; mash bill for, 91–92, 93; non-GMO, 95–96; in oak barrels, 93–94, 99–101; organic farming and, 110; pesticides and, 109–10; technology for, 101–5
wildfires: in California, xiv, 8, 11, 12–13, 49, 76; in Israel, 49, 52, 55
Williams, Adam, 70, 71, 72
wine. *See specific topics*
The Wine Advocate, 10, 20, 122
Wine Enthusiast, 12
The Wine Route of Israel (Montefiore), 40
The Wines of South Africa (Clarke), 157, 173, 174
Wine Spectator, 29, 122; on copper sulfate and lime, 165; on Patagonia, 114, 116–17
Winter Storm Uri, in Texas Hill Country, 133–37, 141–43, 183n1
Wiston Estate, 74
wood barrels. *See* oak barrels
Woodinville, 107–8

Xylellla fastidiosa, 140

Yarrow, Alder, 32–33
Yates, Jessica, 142
Yates, Ron, 134–47, 154
Yates, Tommy Joe, 138, 139, 140
Yellow Label, 80
Young, Brenda, 149
Young, Robert, 149–53
Zandberg, Tamar, 49
Zapata, Nicolas Catena, 125–26
Zikhron Ya'akov, 59
Zuccardi, Sebastian, 119–22, 124, 127, 131

ABOUT THE AUTHOR

Brian Freedman is a wine, spirits, travel, and food writer; restaurant and beverage consultant; wine and spirits educator; and event host and speaker. He regularly contributes to *Food & Wine* digital, Forbes .com, *Whisky Advocate,* and *SevenFifty Daily*; has contributed to *Travel + Leisure* online, *Departures* online, *The Bourbon Review*, and more. He also hosted several wine and spirit pairing segments for the CNN Airport Network. In 2019, Brian was awarded a fellowship to attend the Symposium for Professional Wine Writers in Napa Valley. He consults for restaurants on their beverage programs; regularly hosts virtual and in-person wine and spirit tastings for corporate and private clients; and has traveled to more than fifty countries and territories around the world and extensively throughout the United States to experience the food, drink, and culture for his work. He lives outside of Philadelphia with his wife, Steffi; daughters, Olivia and Sophie; and dog, Murray.

CPSIA information can be obtained
at www.ICGtesting.com
Printed in the USA
BVHW071324141122
651043BV00001B/2